Technical
Writing That Works

Fourth Edition

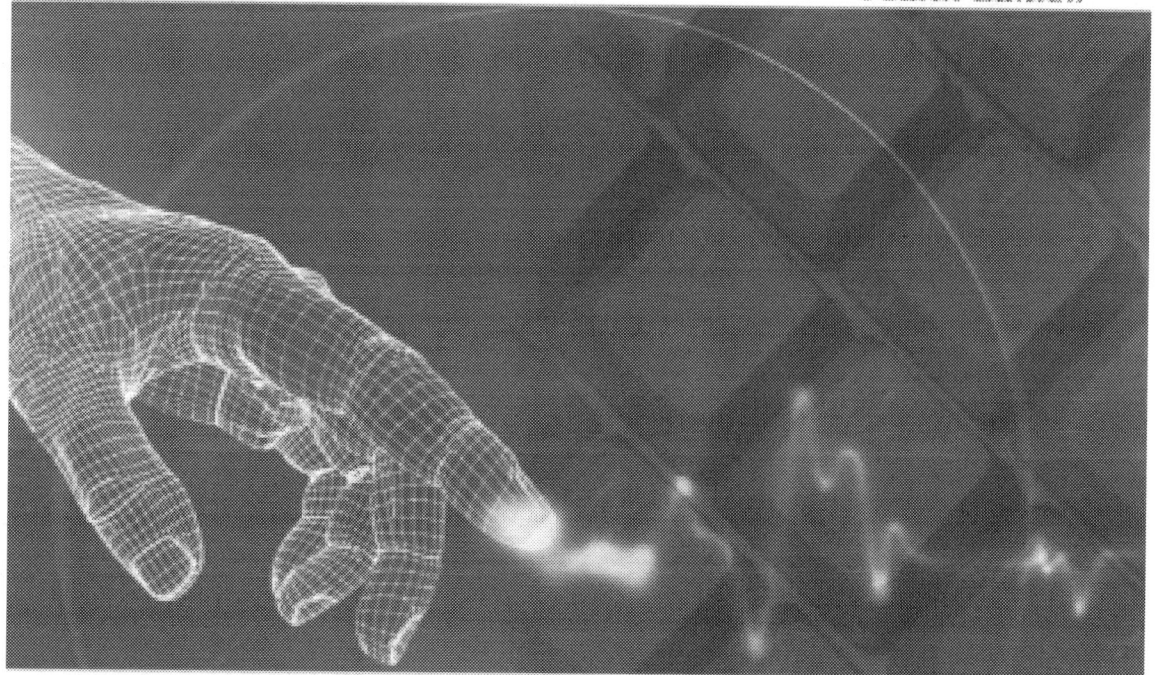

Edward J. Altmann
&
George J. Hallesky, Ed. D.

Technical Writing That Works
Fourth Edition

Edward J. Altmann

and George J. Hallesky, Ed. D.

Johnson College
Scranton, Pennsylvania

authorHOUSE®

AuthorHouse™
1663 Liberty Drive
Bloomington, IN 47403
www.authorhouse.com
Phone: 1-800-839-8640

© *2011 Edward J. Altmann & George J. Hallesky, Ed. D. All rights reserved.*
4th Edition 2012
MLA format on technical research paper edited by Ms. Heather Bonker.

No part of this book may be reproduced, stored in a retrieval system, or transmitted by any means without the written permission of the author.

First published by AuthorHouse   10/18/2011

ISBN: 978-1-4670-6150-6 (sc)
ISBN: 978-1-4670-6149-0 (ebk)

Printed in the United States of America

Any people depicted in stock imagery provided by Thinkstock are models, and such images are being used for illustrative purposes only.
Certain stock imagery © Thinkstock.

This book is printed on acid-free paper.

Because of the dynamic nature of the Internet, any web addresses or links contained in this book may have changed since publication and may no longer be valid. The views expressed in this work are solely those of the author and do not necessarily reflect the views of the publisher, and the publisher hereby disclaims any responsibility for them.

Preface
(What is technical writing?)

Technical writing is informative prose (written in every-day language, opposite of poetry) concerning mechanical or scientific subject matter. Technical writing differs from "creative writing" (poetry, play, essay, short story, and novel) in one major way—each reader should receive the same meaning.

The 60's song "Woolly Bully" by Sam the Sham and the Pharaoh's contains the lyric, "Let's not be l seven, come on and learn to dance." What does "l seven" mean? Visualize "l seven" as "L 7." Bring the "L" and the ""7" together to make a square. Although this is excellent creative writing, we should not direct a carpenter to cut a board "l seven."

Technical writing should be clear, concise (succinct or terse), correct, and complete. A technical writing's paragraph format is often supplemented with illustrations, diagrams, charts, pictures, and other visual aids because most technical topics must be visualized to be understood. When we purchase a product with written instructions for assembly, installation, or use, those instructions are technical writing.

This text's chapters are arranged to mesh with the course and the research paper assignment. The research paper is covered as soon as possible so that students have time to complete the paper as the semester continues. The research paper chapter is immediately followed by the content used in the research paper so that students can include these kinds of writing in their research papers.

Table of Contents

Chapter I : Note Taking 1

Chapter II : Style 6
    Writing in the Introduction, Body, and Conclusion 7
    Defining the introduction, body, and conclusion 7

Chapter III : Meaning 9
    Outline of Types of Meaning 9

Chapter IV : Outlining 16
    Errors in Outlining 16
    Outline Formats 17

Chapter V : The Technical Research Paper 26
    Topic Choice 26
    Suggested Research Paper Materials 27
    Determining What To Research 28
    Format of the Research Paper's Eight Parts 34
    Eight-Step Procedure for Completing the Technical Research Paper ... 38
    Example of Technical Research Paper Preparation 39

Chapter VI : Sentence Definition 57

Chapter VII : Summary 59

Table of Contents

Chapter VIII : Instructions 63
   Travel Instructions 63
   Operational Instructions 64

Chapter IX : Description of a Device 68
   General Description 69
   Specific Description 69

Chapter X : Description of a Process 73

Chapter XI : E-Mail/Business Letters 77
   Appearance 77
   The Letter of Inquiry (Letter) 86
   The Order Letter (E-Mail) 87
   The Claim and Adjustment Letters (E-Mail) 89
   The Sales Letter (E-Mail) 91
   The Letter of Transmittal (E-Mail) 93

Chapter XII : The Personal Résumé 98
   The Personal Inventory 98
   Formats for Résumés 106

Chapter I : Note Taking

Notes are taken during classes, meetings, speeches, demonstrations, seminars, interviews, and research. Before taking notes, sit where you can hear and see well.

Notes are written condensations of important information. Notes are similar to the informative summary (Chapter VII), and the outline format (Chapter IV) is excellent to use. Notes should be neatly rewritten and expanded shortly after they are taken.

When taking notes, quickly write, abbreviate, and sketch. Begin by identifying the circumstance-date, time, location, topic, and speaker. Listen for key statements that indicate important content; such as, "it's important," "remember," "be careful," "let me stress," "I want to emphasize," "don't forget," "always," "never," "I repeat," etc. Pay special attention to blackboard writing, handouts, and visual aids. And group related notes into categories.

A notebook is a well-organized storage facility for all written information relevant to a particular topic. Most supervisors, owners, and professionals (nurses, doctors, lawyers, engineers, salespersons, technicians) keep notes about their work; including times, dates, production, procedures, employees, estimations, orders, receipts, and research.

The Course Notebook

A notebook for a college course has much in common with other types of notes and provides good experience in record keeping. The following is an outline of purposes, materials, format, and content of a course notebook.

I. Purposes
    A. To provide experience in keeping records
    B. To supplement (add to) textbooks
    C. To record lectures
    D. To aid study for quizzes, tests, and assignments

II. Materials
    A. Required
        1. Notebook (A three-ring binder for standard, 8 1/2" X 11", lined composition paper is best because you can add paper, store handouts, and reuse binder. If taking more than one course in a school, use only one notebook and divide it into separate sections for each course. By using only one notebook, you are less likely to forget it, bring the wrong one, or become disorganized.)
        2. Paper (Standard size, 8 1/2" X 11", lined, white composition paper is best.)
        3. Pencil or pen

    B. Suggested
        1. Ruler
        2. Dictionary

  3. Subject and/or topic dividers
  4. Hole punch or pocketed dividers for handouts

III. Format
  A. Class identification (Start each day's notes on a clean sheet of paper. Give each class's date and topic. Number each page.)
  B. Order (Keep all notes in chronological [time] order.)
  C. Style (The outline style is best. It is all right and helpful to combine topic and sentence outline styles as is being done in this outline.)

IV. Content (Divide the content into sections.)
  A. Title page (Identify the course(s), the school, and the student.)

Example Title Page

Course Notebook for ENG 101-English Composition I
Spring Semester of 2012

Jason Hamilton
1565 Hemlock Street
Rochester, New York 14445-0151
Telephone: 716.248.6605

Johnson College
3427 North Main Avenue
Scranton, PA 18508-1495
Telephone: 570.342.6404

  B. Table of Contents (Construct the Table of Contents as the course continues and complete it at the course's end.)

Example Table of Contents Page

Table of Contents

                                                                                  Page

	Page
I. Introduction	1
II. Rewritten Notes	
A. The notebook	2
B. Meaning	
1. Denotative	4
2. Structural	4
3. Contextual	6
4. Connotative	8
C. Technical research paper	
1. Topic choice	9
2. Research	9
3. Parts and format	11

—Continue in same format.—

    C. Rewritten notes (Number each page in Arabic numerals in sequence; for example 1, 2, 3, 4, etc.)

D. Assignments (Restart numbering pages in Arabic numerals in sequence; for example 1, 2, 3, 4, etc.)
E. Handouts (Give dates.)
F. Returned evaluations (Give dates.)
G. Original notes (They are still relevant.)

Example of D.-G.

III. Assignments (Restart page numbering with page 1.)　　　　　　　　　　
    A. Chapter I (Exercise 1)　　　　　　　　　　　　　　　　This Notebook
    B. Chapter III　　　　　　　　　　　　　　　　　　　　　　　　Page

        1. Exercise 1　　　　　　　　　　　　　　　　　　　　　　1
        2. Exercise 2　　　　　　　　　　　　　　　　　　　　　　2
        3. Exercise 3　　　　　　　　　　　　　　　　　　　　　　3
        4. Exercise 4　　　　　　　　　　　　　　　　　　　　　　3
        5. Exercise 5　　　　　　　　　　　　　　　　　　　　　　3

—Continue in same format.—

IV. Handouts (Arrange in chronological order without page numbers)
    A. Course syllabus　　　　　　　　　　　　　　　　　　　　2/20/12
    B. <u>Byte</u> magazine editorial　　　　　　　　　　　　　　　　3/5/12

V. Returned Evaluations (Arrange in chronological order without page numbers)
    A. Quizzes
        1. Meaning quiz　　　　　　　　　　　　　　　　　　　　4/5/12
        2. Sentence definition　　　　　　　　　　　　　　　　　5/1/12

—Continue in same format.—

    B. Tests
        1. Meaning and research paper　　　　　　　　　　　　2/1/12
        2. Outlining, sentence definition, and summary　　　　5/16/12

—Continue in same format.—

    C. Assignments
        1. Research paper
        2. Notebook

—Continue in same format.—

VI. Original Notes (No divisions are needed.)

Exercises for Note Taking

Exercise 1 : The Notebook

Directions: As the semester continues, make a notebook for this course.

Exercise 2 : The Notebook

Directions: As the semester continues, make a notebook for all of your courses.

Chapter II : Style

People have their own styles of doing most things, and writing is one of those things. Consider these recommendations for standardizing your style and then write and rewrite to develop your style.

Before writing, be sure you know your topic's content. This is easier said than done. Research, record, and document information. Be sure you know your purpose for writing. Do you want your reader to know that you did, what he or she must do, or what is done by others? Be sure you know your audience. If the reader is one person, tailor the writing for him or her. If the readers are many, tailor the writing for the average reader, but do not forget the interested apprentices or the seasoned craftsmen. It is a challenge.

The chronological (time) order of a completed writing-title, introduction, body, conclusion-is usually not completed in the same chronological order. The order is more like this: body, introduction, conclusion, and title. This writing order is logical because the writer needs to know the writing's body before he or she can title, introduce, and conclude the body.

The Body

The body is the main, middle, and longest section of content.

Divide the body into its parts using the outline format explained in Chapter IV. Expand the outline into sentences. Group sentences into paragraphs. Change paragraphs when the thought changes. The first sentence of a paragraph is often the thesis sentence, giving the paragraph's main thought to be developed into the rest of the paragraph. Paragraphs are usually fewer than 15 typed lines, and a paragraph can be only one sentence or even only one word as in changing conversation.

Add captions, lists, charts, and illustrations to make the content easy to read, visualize, and understand.

The Introduction

The introduction is 50% of the total writing in importance and in effort. If the reader is lost at the start, his or her reading is over.

The introduction can include as many as five parts. The purpose of your writing will help you to determine the parts to use and those to omit.

1. **attention-getter**-a statement, example, question, or quotation that relates to a topic and gets audience attention without identifying the specific topic

2. **identification of specific topic**-a sentence definition covered in Chapter VI

3. **reason(s) for topic choice**-the reason(s) you chose the topic and the reason(s) for your audience to be interested

4. **documentation of information**-identification of your personal experiences, readings, and interviews relating to the topic

5. **summary of body's contents**-usually a one sentence identification of the body's major divisions of content

The Conclusion

The conclusion should be memorable. It is your last chance to impress your readers.

The conclusion's content depends upon the purpose of your writing's body. If you informed, the conclusion might summarize the body's main content. If you have researched, the conclusion might be a recommendation or an explanation of the topic's future. If you have written a business letter, the conclusion might be a statement of appreciation.

The Title

The title is the first chance to make a good impression on the reader. He or she decides whether or not to read your writing because of its title.

Use verbs in titles to add motion and life. Alliteration, the repetition of initial sounds in neighboring words, is often pleasing in a title. Punctuate the title conventionally and do not underline or quote it. Capitalize the first letter of the first and last words. Capitalize all other words except articles (a, an, the), prepositions (about, in, of, etc.), and conjunctions (and, or, neither/nor, etc.) Consider these examples:

Terminate Termites.
Where Will an Underwater Welder Work?
Consider a Home in a Hill.
The Team That Races to the Racer's Rescue
Morrison Motion Detector Detects More.

Avoid titles that are general, misleading, or trite. Consider these examples:

Wheels and Tires
Learning To Be a Mason
Medicine or Miracle?

Writing in the Introduction, Body, and Conclusion

Be clear, correct, concise, and complete. Chapter 7 of Sentence Analysis (Copyright 2005,1999 by Pearson Custom Publishing) includes additional explanations, examples, and exercises.

Be clear. Turn people on by writing clearly.

1. Use quality materials: paper, type, cover, etc.
2. Don't crowd content or waste space.
3. Divide the writing into inviting sections.
4. Include a Table of Contents for long writings.
5. Tell what is.
6. Use small, simple words.
7. Use verbs.
8. Make professional illustrations.

Be concise. Make every word work.

1. Avoid redundant, wordy, or trite phrases.
2. Avoid adjectives and adverbs.
3. Use active voice verbs.
4. Combine similar ideas.
5. Don't tell the reader what he or she knows.

Be correct. Write about what you know well.

1. Identify your purpose and audience.
2. Do research.
3. Verify all information.
4. Use proper grammar, syntax, and mechanics.
5. Read your writing aloud.

Be complete. Imagine yourself as the reader.

1. Include all needed parts from title page to glossary.
2. Spell out acronyms and abbreviations the first time they are used.
3. Define technical terms and jargon.
4. Call attention to cautions and special hints.
5. Include personal experiences, including those recalling early errors.
6. Document all information.
7. Tell the reader where to get more information.
8. Choose a format and stick to it.
9. Make and keep copies of written transactions.

Chapter III : Meaning

The English language has the largest vocabulary of all major languages. There are more than 800,000 English words, and many have multiple meanings. Words do not have meaning in themselves; they are assigned meaning by people, and that meaning can change. A dictionary published in the 1960's would define "deck" as a noun meaning 1. a floor on a ship 2. a roofless platform attached to a house 3. a pack of playing cards, and as a verb 1. to clothe in finery 2. to decorate 3. to knock down. However, a dictionary published after the 1960's would list the previous definitions and a fourth noun usage "a tape deck."

To understand a word's meaning, we must evaluate the word according to four types of meaning: denotative meaning, structural meaning, contextual meaning, and connotative meaning. All words have structural and contextual meanings, and most have varying degrees of the other two types. Some words are heavily denoted, but others are heavily connoted. Dictionaries give denotative, structural, and contextual meanings, but connotative meaning can be only approximated. Additionally, dictionaries include the syllabications, pronunciations, and etymologies (word histories) of their words.

> Example: **screw•dri•ver** (skroo dri v r) n. 1. A hand tool used to turn screws. 2. A cocktail made with orange juice and vodka. [ME *skrewe* < Ofr. *escrove*, female screw or nut]
>
> Explanation: The word "screwdriver" has three syllables. It is pronounced as it is phonetically written in parentheses. The lower case letter "n." means the word is a noun. The word denotes two meanings depending upon its context. And, the word's history is traced to Middle English preceded by Old French in which "screw" meant "female screw or nut."
>
> Example sentence: Before drinking, the electrician used his screwdriver to stir his screwdriver.

Outline of Types of Meaning

I. Denotative Meaning
   A. Identification-A heavily denoted word shows a strong word to object relationship.
   B. Qualities
      1. Concreteness-Concrete nouns naming persons, places, and things are heavily denoted.
      2. Primitiveness-As children, we learn heavily denoted words first so that we associate the person (Mom), the place (home), or the thing (water) with each word's sound.
      3. Sameness-Heavily denoted words mean the same things to everybody

C. Example-If a draftsperson, a highway construction worker, a machinist, a musician, a baseball player, or a housewife were asked to define the following heavily denoted words, their definitions would be the same: <u>carpenter</u>, <u>nurse</u>, <u>lawyer</u>, <u>house</u>, <u>factory</u>, <u>stadium</u>, <u>refrigerator</u>, <u>television</u>, <u>automobile</u>, <u>minute</u>, <u>year</u>, and m<u>ile</u>.

II. Structural Meaning

A. Identification-Structural meaning shows a sign to sign relationship. The "signs" are grammatical or mechanical. The grammatical signs can be evaluated from the spoken word, phrase, or sentence. But, the mechanical signs can be evaluated only when the word, phrase, or sentence is written.

B. Grammatical signs
    1. Part of speech (noun, pronoun, verb, etc.)
    2. Usage (subject, direct object, predicate noun, etc.)
    3. Number (singular or plural)
    4. Agreement (subject and verb, pronoun and antecedent, adjective and noun, etc.)

C. Mechanical signs
    1. Spelling
    2. Capitalization
    3. Punctuation

D. Example sentence-Most zats have dails. What do "zats," and "dails" mean? Most people would say that the words have no meaning because they are not in the dictionary. They have no denotative meaning, but when used in a sentence, they do have structural meaning.
    1. Criteria

	zats	dails
a. Grammatical signs		
1) Part of speech	noun	noun
2) Usage	subject	direct object
3) Number	plural	plural
4) Agreement	agrees with plural verb "have" and with plural adjective "Most."	N/A
b. Mechanical signs		
1) Spelling	acceptable (has vowel and is pronounceable)	acceptable (has vowel and is pronounceable)

    2) Capitalization   common noun   common noun
    3) Punctuation   none   none

  2. Conclusion-Therefore, a nonsense word has no denotative (dictionary) meaning, but when a nonsense word is used in a sentence, the nonsense word will have structural meaning.

 E. Meaning affected by mechanical signs
  1. Spelling
   a. Spelling is (impotent, important) to meaning.
   b. Jackson Auto Parts has its parts (filled, filed) in computers.
   c. That's a (stationary, stationery) cabinet.
   d. There are three (to's, too's, two's) in the English language. (All are incorrect spellings; the meaning can't be spelled correctly.)
  2. Capitalization
   a. That wax is (polish, Polish).
   b. The concert featured (the, The) Grateful Dead and (the, The) Who.
   c. The Rockport Technical College will have (cars on campus, Cars on Campus) on Sunday, September 12th.
   d. Frank is (frank, Frank).
  3. Punctuation
   a. The driver said, "The flagman made the mistake." "The driver," said flagman, "made the mistake."
   b. The architect's designs reflected the homeowner's wishes.
    The architects' designs reflected the homeowners' wishes.
    The architect's designs reflected the homeowners' wishes.
    The architects' designs reflected the homeowner's wishes.
   c. Start with 15 matches, take six away, and be left with "TEN."
   d. Railroad Crossing, Look Out for the trains—can you spell "it" without any r's?

III. Contextual Meaning
 A. Identification-Contextual meaning shows a word to an object to a sign relationship. It is a combination, a hybrid, of denotative and structural meanings. A heavily contextual word gets its specific meaning from the other words accompanying it in the phrase or sentence. Used alone, a heavily contextual word has all of the possible meanings that dictionaries list for it.

 B. Example paragraph-Most zats have dails. If the parent zats have dails, then a Litter of their zats will also have dails. There is only one breed of domestic house zat that does not have a dail. It is called a Manx and is similar to a bob zat and a lynx.

"Zats" and "dails" are still nonsense words without denotative (dictionary) meaning, but now they have contextual meaning and structural meaning. "Zats" are similar to cats, and "dails" are similar to tails. We know this because the denotative and structural meanings of the other words tell us.

C. Example word-If people of various occupations were asked to define the word "pitch," their definitions would vary.
1. a draftsperson-the angle of a roof
2. a highway construction worker-the black, tarry substance
3. a machinist-the slant of a thread
4. a musician-a quality of sound
5. a baseball player-the delivery of a baseball by a pitcher
6. a housewife-a sales pitch

D. Example word-Notice how the context of the word "run" in phrases affects its meaning.
1. to run a mile (to travel on foot faster than a walk)
2. to score a run (a point in baseball)
3. a run in the paint (a drip)
4. to run a machine (to operate)
5. to run with a bad crowd (to associate with)
6. a run in a stocking (an unraveling flaw)
7. to run for office (to attempt to be elected)
8. a ski run (a path)
9. a run of ore (a vein)
10. the rise and the run of stairs (the depth of thread)
11. a run-of-the-mill day (an average day)
12. the run of the place (complete freedom)
13. the run of production (the amount of)
14. a run on gasoline (a high demand)
15. to run the rack (to make consecutive pool shots)
16. a run in with the police (a confrontation)
17. the running water (flowing)
18. a run-on sentence (one continuing into the next)
19. a propeller run (the distance of blade rotation)
20. the runs (diarrhea)

IV. Connotative Meaning
Identification-Connotative meaning shows a word, to a quality, to a person relationship. The person aspect results from each individual's attitudes, background, and experiences that are never the same for any two people. Therefore, heavily connoted word never has exactly the same degree of meaning to all people or even to the same person. We have all said, "I was scared." But, our degrees of

being scared were always different. Abstract nouns naming qualities and ideas are heavily connotated as are many adjectives and adverbs. For example: Many smart liberals are overly nervous about the spirit of nationalism.

B. Example word-"old"
1. Denotative meaning: (word to quality) a measurement of age, the opposite of "young"
2. Connotative meaning: (word to quality to person) depends upon the age of the user and the noun he or she is describing
   a. The 10 year old said, He is old." (older than 10 yrs.)
   b. The 20 year old said, "He is old." (older than 20 yrs.)
   c. The 40 year old said, "He is old." (older than 40 yrs.)
   d. The gnat is old. (24 hours)
   e. The dog is old. (15 yrs.)
   f. The parrot is old. (50 yrs.)
   g. The Galapagos turtle is old. (200 yrs.)
   h. The Parthenon is old. (1000 yrs.)

C. Quotations expressing connotative meaning
1. "A house is not a home."
2. "Beauty is in the eye of the beholder."
3. "Happiness is different things to different people."
4. "One man's trash is another's treasure."
5. "There's a thin line between love and hate."

D. Range of connotation (Each group denotes the same thing or quality, but each connotes a different level of meaning. A thesaurus is a book of words and their synonyms similar to these examples.)

Favorable	Neutral	Unfavorable
normal	average	mediocre
officer	policeman	cop
unique, exotic	different weird,	strange, bizarre
replica	copy, duplicate	fake, counterfeit
scholar	student, pupil	bookworm

V. Conclusion

When interpreting information, evaluate its meaning in relation to the four types of meaning. When speaking or writing, choose terms that conform to the four aspects of meaning, use heavily denoted terms rather than heavily connoted words, and avoid adjectives and adverbs.

Exercises for Meaning

Exercise 1 : Denotation and Connotation

Directions: Identify the following words as heavily denotated (D) or heavily connotated (C):

_____ 1. intelligent   _____ 2. smart   _____ 3. 150 IQ
_____ 4. expensive   _____ 5. $1,000   _____ 6. valuable
_____ 7. short   _____ 8. one block   _____ 9. one inch
_____ 10. one mile   _____ 11. 50 knots   _____ 12. fast
_____ 13. tall   _____ 14. 100 stories _____ 15. slow
_____ 16. 100 rpm   _____ 17. 1000 lbs.   _____ 18. 1 oz.
_____ 19. heavy   _____ 20. thin   _____ 21. 1 m.m.
_____ 22. today   _____ 23. soon   _____ 24. now
_____ 25. hot   _____ 26. 2000°   _____ 27. cold
_____ 28. far   _____ 29. close   _____ 30. near

Exercise 2 : Structural Meaning

Directions: Answer these structural questions about the underlined nonsense words: grammar (part of speech, usage, number, and agreement) and mechanics (spelling, capitalization, and punctuation). Some might not apply.

1. The <u>wickment</u> is a <u>plunage</u> of materials.

2. All <u>Rogents</u>® has <u>gendadic</u> transmissions.
3. We <u>gndts</u> to the <u>binderhood</u>.
4. The foreman <u>juntly</u> fired the <u>splentons</u>.
5. Being <u>terculated</u>, <u>reciplate</u> your goals.

Exercise 3 : Contextual Meaning

Directions: Use these heavily contextual words as nouns in various phrases or sentences to show their different meanings.

1. hammer   2. nail   3. figure
4. lead   5. hand   6. nut
7. bolt   8. stock   9. line
10. heat

Exercise 4 : Connotative Meaning

Directions: Arrange the following terms from most unfavorable connotation to most favorable connotation: occupation, work, calling, job, profession, employment, career, livelihood, bread and butter, vocation, avocation.

Exercise 5 : Range of Connotation

Directions: Make a list of favorable and unfavorable terms that denote the same neutral qualities listed.

1. angry
2. tolerant
3. calm
4. Mother and Father
5. partner
6. cautious
7. nervous
8. tired
9. happy
10. confident

Chapter IV : Outlining

The ability to outline is the ability to think logically. People outline all of the time. Suppose that you want to attend a sporting event 150 miles from your home. Before leaving, you plan how to dress, how to pay, and how to travel. You then divide the previous three major topics: how to dress (hat, shirt, jacket, pants, and shoes), how to pay (transportation, admission, and food), and how to travel (directions, vehicle, and transfers).

Whenever you take on a service, repair, or construction task, you ask yourself, "Do I have the skill, equipment, and material needed?" Each of these three major categories is divided into its divisions: skill (experience, research, and subcontracting), equipment (hand tools, power tools, measuring instruments, and test taking devices), and material (all parts and their payments).

There are two types of outlines, classification and partition. In classification a plural topic is divided into its types (styles of homes, chapters of books, kinds of machinery used in a machine shop, types of electric motors, welding processes, methods of treating a disease, etc.) In partition a singular topic is separated into its parts (house, tire, hypodermic needle, lathe, application form, electric drill, micrometer, etc.).

Errors in Outlining

Error #1 : Content is not coordinate.

The content of an outline on any level must be coordinate or parallel in content and in grammatical structure. If home heating systems were classified according to their fuel as the coal, fuel oil, natural gas, electricity, wood stoves, solar energy, and geothermal, their divisions would not be coordinate in either content or structure. "Wood stoves" is not a fuel, and the adjective "geothermal" is not a noun. Changing "wood stoves" to "wood" and "geothermal" to "geothermal steam" would be correct.

Remember that each level must be coordinate, not the whole outline. Major divisions could be nouns. One noun could be divided into infinitive phrases, another into adverb clause, and still another into nouns.

Error #2 : Content is overlapping.

Each item in an outline should be capable of being placed into only one category. If automobiles were classified according to fuel mileage, the breakdown into those getting 0-15 mpg, those getting 15-30 mpg, and those getting 30 mpg or more would overlap. Cars getting exactly 15 mpg or 30 mpg could be put into two categories. A correct version is those getting fewer than 15 mpg, those getting 15-30 mpg, and those getting more than 30 mpg.

Error #3 : Content is not mutually exclusive.

Each division in an outline should make sense even if the other divisions were omitted. If vats were classified according to capacity as vats in group I, vats holding fewer gallons than group I, and vats holding more gallons than group I, the divisions would not be mutually exclusive because they all depend upon group I for meaning and could not exist alone. An improvement is vats holding fewer than 50 gallons, vats holding 50-100 gallons, and vats holding more than 100 gallons.

Error #4 : Content is incomplete.

Each division of an outline should be complete. If wrenches were classified according to types as adjustable wrenches, combination wrenches, and ratchet wrenches, the divisions would be incomplete without adding "allen wrenches" and "torx wrenches."

Outline Formats

1. **Use either a topic outline** in which each entry is a word, phrase, or dependent clause, or **a sentence outline** in which each entry is a complete sentence. We will concentrate on the topic outline.

2. **Use the traditional number/letter format.** The major divisions of the title are represented by Roman numerals followed by periods. Second-level divisions are represented by capital letters followed by periods, third-level divisions are represented by Arabic numerals followed by periods, fourth-level divisions are represented by lower case letters followed by periods, fifth-level divisions are represented by an Arabic numerals followed by one parenthesis, and sixth-level divisions are represented by lower case letters followed by one parenthesis.

3. **Determine the importance and the intended audience of your outline.** The same content can be outlined more than one correct way. Imagine that you are a new-car salesperson at a dealership selling many brands and styles of automobiles. One customer asks you which models are least expensive, a second asks which are the most luxurious, a third asks which get the best fuel mileage, a fourth asks which are the safest, etc. The models you sell are the same, but each customer (intended audience) has a different purpose (a different order for his or her outline).

4. **Choose appropriate orders for the outline's divisions:** chronological or time order as in listings a task's steps, spatial order as in describing a house from basement to attic, familiarity order as in identifying engine fuels from gasoline to solar energy, and importance order as in evaluating a person's physical condition from vital signs to appearance.

5. **Indent each of an outline's levels approximately 1/2" more than the previous level.**
Keep all corresponding levels in a straight line regardless of where they are used.

6. **Capitalize the outline's title and its content representing Roman numerals;** except articles, prepositions, and conjunctions that don't start or end the level's content.

7. **Capitalize the first word of all five other divisions, as well as their proper nouns and adjectives.**

8. **Place a period after the numbers and letters of the first four levels of divisions.** Place one parenthesis after the 5th and 6th levels of divisions.

9. **Be sure that each level of an outline has at least two parts to its division, or the previous number or letter has not been divided.**

10. **Skip a line before each Roman numeral; otherwise, single space.**

Note: Classification and partition can be use in the same outline but only in different levels of the outline.

Example of Topic Outline Format
(Countless variations are possible.)
Title of Outline

I. First Major Division of Outline or Title

II. Second Major Division of Outline or Title
   A. First division of Roman numeral II
   B. Second division of Roman numeral II

III. Third Major Division of Outline or Title
   A. First division of Roman numeral III
   B. Second division of Roman numeral III
     1. First division of capital letter B
     2. Second division of capital letter B

IV. Fourth Major Division of Outline or Title
   A. First division of Roman numeral IV
   B. Second division of Roman numeral IV
     1. First division of capital letter B
     2. Second division of capital letter B
     3. Third division of capital letter B
       a. First division of Arabic numeral 3
       b. Second division of Arabic numeral 3
         1) First division of lower case letter b
         2) Second division of lower case letter b
       c. Third division of Arabic numeral 3
     4. Fourth division of capital letter B
   C. Third division of Roman numeral IV
     1. First division of capital letter C
     2. Second division of capital letter C

             a. First division of Arabic numeral 2
             b. Second division of Arabic numeral 2
                1) First division of lower case letter b
                2) Second division of lower case letter b
                   a) First division of Arabic numeral 2 and parenthesis
                   b) Second division of Arabic numeral 2 and parenthesis

V. Fifth Major Division of Outline or Title

Topic Outline Classifying American Automobile Tires

I. Tread Design
- A. Summer
- B. Winter
- C. All-season
- D. Recreational

II. Wheel Size (Diameter of Wheel)
- A. 12"
- B. 13"
- C. 14"
- D. 15"
- E. 16"
- F. 17"
- G. 20"

III. Profile (Distance from Tread to Wheel)
- A. 78 series
- B. 70 series
- C. 60 series
- D. 50 series
- E. 40 series

IV. Width
- A. P 155
- B. P 165
- C. P 175
- D. P 185
- E. P 195
- F. P 205
- G. P 215
- H. P 225
- I. P 235

V. Construction
    A. Bias-ply
    B. Radial-ply
    C. Belted bias-ply or radial
        1. Steel
        2. Fiberglass
        3. Steel and fiberglass

VI. Sidewall Design
    A. Blackwall
    B. Whitewall

VII. Manufacturer
    A. Domestic
        1. Goodyear®
        2. Firestone®
        3. Etc.
    B. Foreign
        1. Bridgestone®
        2. Michelin®
        3. Etc.

VIII. Price
    A. Inexpensive-under $50
    B. Average-$50 to $100
    C. Expensive-over $100

IX. Inflation Method
    A. Tubeless
    B. Tubed

X. Miscellaneous Ratings
    A. Mileage
    B. Inflation poundage
    C. Traction
    D. Handling
    E. Guarantee

Note: Consider the tires on the car you drive most often. Those tires will fit into each of the preceding 10 categories.

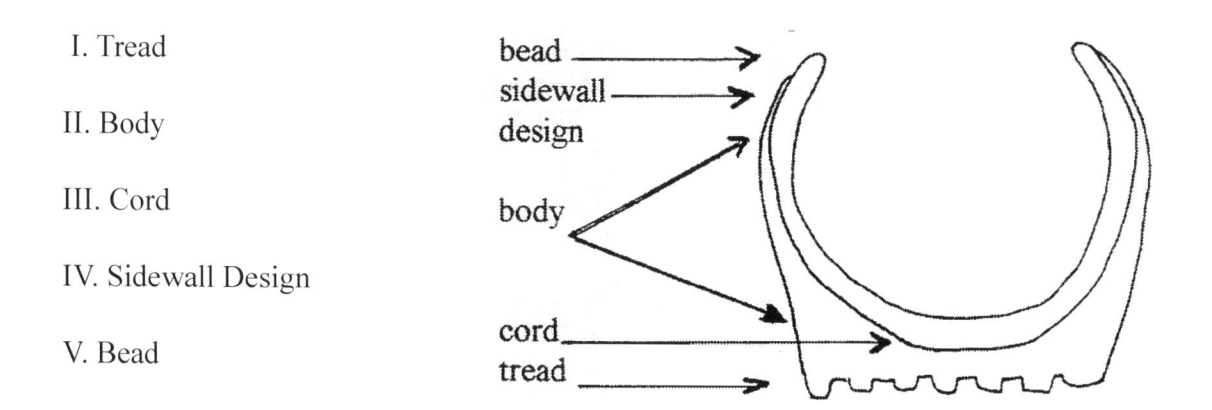

Topic Outline Partitioning an Automobile Tire

I. Tread

II. Body

III. Cord

IV. Sidewall Design

V. Bead

Topic Outline Partitioning an Automobile Tire and Classifying the Divisions of Its Parts

I. Tread
    A. Summer
    B. Winter
    C. All-season
    D. Recreational

II. Cord and Body
    A. Bias-ply
    B. Radial-ply
    C. Belted bias-ply or radial
        1. Steel
        2. Fiberglass
        3. Steel and fiberglass

III. Sidewall Design
    A. Blackwall
    B. Whitewall

IV. Bead

Note: Check the topic outlines used in this text in Chapters I, III, V, & XI.

Exercises for Outlining

Exercise 1 : Classification

Directions: Classify the following: titles into Roman numerals, the titles' major divisions: Automobile Fluids; Hammers; Telephones; Welding Processes; Applications; Home Styles; Primary Colors; Electronic Audio Communication Devices; Floor Coverings; Fences; Parts of Speech; Machinery of a Machine Shop, Carpentry Shop, Auto Shop, etc.

Exercise 2 : Partition

Directions: Partition the following: titles into Roman numerals, the titles' major divisions: Nail; Arrow; Battery-Powered Flashlight; Gauges of a Car's Dashboard; Set of Stairs; Book; Two-Story House; Fireplace; Job Application; Outdoor Deck; A Tool Used in Electrical Shops, Welding Shops, Drafting Offices, etc.

Exercise 3 : Classification and Partition

Directions: Classify the Roman numerals into capital letters.

Saws

I. Man-Powered
   A. For ?
   B. For ?
   C. For ?
   D. For ?

II. Fuel-Powered
   A. For ?
   B. For ?
   C. For ?
   D. For ?

Directions: Classify the Roman numeral <u>II</u>. and capital letter <u>C</u>.

Home Building Materials

I. Wood

II. Metal
   A. ?
   B. ?

III. Plastic

IV. Glass

V. Masonry
   A. Cement
   B. Natural Stone
   C. Blocks
      1. ?
      2. ?
      3. ?

Directions: Classify the Roman numeral <u>III</u>., its capital letters <u>A</u>. and <u>B</u>., and its Arabic numeral <u>2</u>.

China Cabinet Materials

I. Lumber

II. Glass

III. ?
   A. ?
      1. Glue
      2. Metal
         a. ?
         b. ?
   B. ?
      1. Light
      2. Hinges
      3. Pulls

IV. Stain and Sealer

Directions: Partition the Roman number <u>V.</u> into capital letters.

AM-FM Portable Radio

I. Housing

II. Receiving Module

III. Speaker (s)

IV. Batteries

V. Controls
    A. ?
    B. ?
    C. ?
    D. ?

Directions: Partition the capital letter <u>B.</u> and the Arabic numeral <u>4.</u>

Building a Piece of Furniture

I. Design

II. Fabricating
    A. Cutting
    B. Assembling
        1. ?
        2. ?
        3. ?
        4. Finishing
            a. ?
            b. ?
            c. ?

Exercise 4 : Classification

Directions: Use 20 of the following 24 divisions of content to make an outline classifying the topic Methods of Transportation: 1. bicycle 2. automobile 3. animal-powered 4. air mail 5. stagecoach 6. airplane 7. fuel-powered 8. canoe 9. pigeon 10. dogsled 11. jet propulsion type 12. human-powered 13. motorcycle 14. horseback 15. propeller propulsion type 16. train,17. sports car (two passenger)18. family car (two to six passenger) 19. skis (cross-country) 20. ice skates 21. Traveling and sightseeing 22. ocean liner (ship) 23. stationary 24. Roller skates

Exercise 5 : Classification and Partition

Directions: Make a detailed topic outline including as many subdivisions As appropriate for the following: An Interstate Highway System, a Professional Baseball Stadium, Styles of Professional Baseball Stadiums, Automobile Engine Performance, Types of Fuel, Step in Re-roofing a House, Detailing a Car, Examining an Animal, Major Systems of a Car, Anatomy of a Dog, Cat, Bird, etc., or an Exercise 1 example.

Exercise 6 : Classification and Partition

Directions: Compose an estimation form for a construction, repair, or service project that applies to your career. Include company identification, customer identification, date of estimate, description of job, materials needed, labor explanation, and total cost.

Exercise 7 : Outlining a Completed Writing

Directions: Construct a topic outline partitioning the topic of Body Language as it is represented in the following writing.

Body Language

Language is oral or written communication, and body language is communication expressed in physical mannerisms that show a person's attitudes, emotions, feelings, and moods. Body language can be as important as oral or written language. We use body language in appearance, dress, and gestures.

If we use a job interview as an example situation, the interviewer will closely evaluate the interviewee's body language. Suppose that you are the personnel manager and your office overlooks the visitors' parking lot. Your first interview is scheduled for 10:00 a.m. You will actually start the interview as soon as you see the applicant pull into the parking lot. How early is he or she? What is the condition of his or her car? If you can see into the car, are there coffee cups all over? Is the applicant finishing dressing in the car such as putting on a necktie, cologne, or perfume? As the applicant enters the building, does he or she wipe his or her feet? These are all body language that the applicant used probably unknowingly.

When the applicant enters the waiting room, does the applicant introduce himself or herself to the secretary? Does the applicant sit before being directed? If literature were available in the waiting room, did the applicant read company literature or People magazine? Is the applicant chewing gum or smoking? Some secretaries to employment managers complete a written evaluation of a job applicant's behavior prior to the interview. Again the applicant might not be aware of his or her body language and its effect upon employment.

As the applicant enters the interviewer's office, does the applicant introduce himself or herself, use the interviewer's surname titled with respect ("Mr.," "Mrs." etc.), and shake hands? How is the applicant dressed—conservative, flashy, sloppy? How is the applicant groomed—hairstyle, shave, cleanliness, fingernails? How does the applicant sit—erect, slouched, fidgety? How does the applicant use eye contact—direct, glancing, none? Some nervousness is expected, but a cocky or overconfident attitude is unacceptable.

When the interview ends, does the applicant thank his or her interviewer and the interviewer's secretary? When entering his or her car, does the applicant remove a necktie, pull a shirt from pants, or drive away carelessly? Still, he or she is being evaluated through body language.

Body language is a big part of an employment interview, and often the interviewee is unaware of it.

Chapter V : The Technical Research Paper

The technical research paper is a formal writing telling what its writer has learned about a mechanical or scientific topic. It is similar to a term paper.

We must cover this paper early so that required research papers can be completed before the course ends. The drawback to early coverage is that some of the aspects of writing the research paper are covered after it. The appropriate chapters are covered as soon as possible after this one.

The format of any formal paper is largely determined by its writer. In this chapter you are expected to follow the suggested format. This format is complete, having eight parts. The parts are required so that you will consider using these parts in your future writing assignments. Then you decide what parts to include, omit, modify, or expand.

This chapter covers choosing a topic, determining what to research, locating research materials, formatting the nine parts, and completing the final paper.

"Research" in its purest sense suggests finding something which no one else knows (a cure for AIDS). In reference to your technical research paper, it means finding something that others do know but that you do not.

Topics, lengths, due-dates, resources, and content are determined by your instructor.

Topic Choice

If a topic is not assigned to you, choose one that is mechanical or scientific, is related to your technical studies, and is interesting to you. Your personal interest is most important, but do not choose a topic that you already know. Usually, we can research an unfamiliar aspect of an otherwise-understood topic. Thus, a carpenter who hunts might research trap, bow, or rifle stock construction.

Be sure that your topic is specific enough to be researched in a month and to be explained completely in your instructor's maximum wording limitations. "Houses" is much too large, "A-Frame Houses" is too large, but "Framing an A-Frame House" is fine.

Be sure information is available. The President of the United State's limousine would be an interesting topic for an automotive student, but are its special features (defense, weapons, communications) classified information?

Example Topics

Architectural Drafting
solar heating
underground housing
green construction
landscaping
handicap access

Automotive
Brinks truck
synthetic motor oil
turbocharging
Wankle engine
anti-lock brakes

Biomedical
plasma banks
organ transplants
life-flight aviation
ambulances
electronic monitors

Example Topics (Cont.)

Carpentry	**Computer**	**Diesel**
termites	motherboards	air brake systems
boat construction	firewalls	fuel injection systems
furniture restoration	networking systems	diesel engines
green construction	servers	diesel tune-up procedures
concrete sidewalks	hardware/software	computerization of diesel

Distribution	**Electrical**	**Electronic**
materials handling	residential wiring	digital electronics
international logistics	commercial wiring	semiconductors
computer operations	PLC programming	PLCs
inventory control	industrial mechanics	semiconductors
shipment methods	electric automobiles	vacuum tube

Precision Machining	**Radiologic**	**Veterinary Science**
Jewelry making	alarm systems	Manx cat
blacksmithing	lasers	whelping
gunsmithing	radar	deer tick
heat treating	fax machines	African gray parrot
chroming	electric automobiles	rabies

Welding
underwater welding
nondestructive testing
woodburner construction
bluing
blacksmithing

Suggested Research Paper Materials

I. Script
    A. Typing is professional and preferable.
    B. Use only one side of the page.
    C. Allow a 1" to 1 1/2" margin around the pages. Keep consistent margins on all pages.

II. Paper
    Use unlined, white paper 8 ½" X 11"

III. Cover
    No cover is required. Staple finished paper in the upper left-hand corner.

Determining What To Research

After choosing your topic, write a list of questions for which you will try to find answers. Three to five questions should be answered in the paper's body. These questions will also help you to narrow your topic. You may find that one question includes from three to five divisions and can become the whole topic.

>Example
>student's major : carpentry
>specific topic : termites
>specific questions to research :
>>1. What is the life cycle of termites?
>>2. Why do houses become infested with termites?
>>3. How are termites discovered?
>>4. How are termites exterminated?
>>5. Can termite damage be repaired?
>>6. Can homes be protected against termite infestation?

You may find so much information on termite extermination that it becomes your topic divided into (1) Do-it-yourself extermination (2) professional extermination services (3) prevention of re-infestation.

Conducting Research

After choosing a topic and determining the questions to research, you are ready to research. The library is the most useful facility for valid research materials; many of these materials may be accessed online, but visits to working examples, interviews, writing or phoning requests, and online sources (provided they are validated) are also valuable sources of information.

The Library

No two libraries are exactly the same in layout, materials, and indexing. Here we will review the materials and indexing that are common to most libraries; such as, the reference works, the card catalog, and magazine indexes. Electronic indexing systems are best learned at individual libraries. Librarians and written usage explanations are always available to help you. Most libraries today are almost entirely computerized.

Reference Works

Most libraries have a section called the "Close Reserve," which contains the reference works that can not be checked out of the library. Encyclopedias, dictionaries, handbooks, almanacs, bibliographies, and yearbooks are the typical general reference works.

Card Catalog

Libraries organize their books according to one of two classifications, the Library of Congress Classification System or the Dewey Decimal Classification. The Library of Congress system uses 20 letters of the alphabet to divide library books into 20 categories of subject-matter as follows:

      A - General reference works
      B - Philosophy, psychology, religion
      C - History, auxiliary sciences
      D - History and topography (except America's)
E and F - American history
      G - Geography, anthropology
      H - Social sciences
      J - Political science
      K - Law
      L - Education
      M - Music
      N - Fine arts
      P - Language and literature
      Q - Science
      R - Medicine
      S - Agriculture, plant and animal industry
      T - Technology
      U - Military science
      V - Naval science
      Z - Bibliography, library science

The Dewey Decimal system, the more commonly used system, uses numbers to divide its books into ten categories of subject-matter as follows:

      000 - General reference works
      100 - Philosophy and related disciplines
      200 - Religion
      300 - Social sciences
      400 - Language
      500 - Pure sciences
      600 - Technology (Applied sciences)
      700 - Arts
      800 - Literature
      900 - General geography, biography, history

These basic groups are divided and subdivided numerous times. In addition to the general reference works in the 000 section, students of technical subjects most often use the books of the 500, 600, and 700 groups. Their major subdivisions are as follows:

500 - Pure sciences
510 - Mathematics
520 - Astronomy and allied sciences
530 - Physics
540 - Chemistry and allied sciences
550 - Sciences of earth and other worlds
560 - Paleontology
570 - Life sciences
580 - Botanical sciences
590 - Zoological sciences
600 - Technology (Applied sciences)
610 - Medical sciences
620 - Engineering and allied operations
630 - Agriculture and related technologies
640 - Home economics and family living
650 - Management and auxiliary services
660 - Chemical and related technologies
670 - Manufacturers
680 - Manufacture for specific uses
690 - Buildings

700 - Arts
710 - Civic and landscape art
720 - Architecture
730 - Plastic arts and sculpture
740 - Drawing, decorative, and minor arts
750 - Painting and paintings
760 - Graphic arts and prints
770 - Photography and photographs
780 - Music
790 - Recreational and performing arts

Specific books within the library classifications are represented by three identification cards in the card catalog, which is an alphabetical arrangement of 3" X 5" cards according to author, title, and subject matter. Computerized libraries have these three cards in an electronic database to list the holdings of the particular library. Therefore, each library book is listed on each type of card as these examples show:

Author Card

574.19
Fo
        Forney, Lois W.
        Chemical Principles for Life/Lois W. Forney
        Englewood Cliffs, N.J. : Prentice-Hall, c. 1978.
          xv, 574 p. : ill. ; 24 cm.
          includes index
          ISBN 0-13-128694-3 : $12.95

Title Card

574.19
Fo
        Chemical Principles for Life
        Forney, Lois W.
        Englewood Cliffs, N.J. : Prentice-Hall, c. 1978.
          xv, 574 p. : ill. ; 24 cm.
          includes index
          ISBN 0-13-128694-3 : $12.95

Subject Card

574.19
Fo
        Chemistry
        Forney, Lois W.
        Chemical Principles for Life/Lois W. Forney
        Englewood Cliffs, N.J. : Prentice-Hall, c. 1978.
          xv, 574 p. : ill. ; 24 cm.
          ISBN 0-13-128694-3 : $12.95

    After locating a book in the card catalog, you use the book's call number 574.19 Fo to find the book in the library's stacks of books. The call number is also taped to the book's bottom binding. After recording the book's call number, refer to the library's floor plan showing the positioning of books in the library's classification system. You should then be able to find the book if it has not been checked out. The librarian can inform you if the book has been checked out and when it is due to be returned. Many libraries will also allow you to put a hold on the book to keep it for you when it is returned and to notify you of its return.

Magazine Indexes (Also available online in most libraries)

The Readers' Guide to Periodical Literature and the Applied Science and Technology Index are the two major indexes to magazine articles. The Readers' Guide indexes over 240 popular magazines; such as, Time, Newsweek, Sports Illustrated, Scientific American, Popular Science, etc. The Applied Science and Technology Index indexes over 800 specialized technology periodicals; such as, Iron Age, Compressed Air, Byte, Data Processing, etc. Both indexes are published monthly, quarterly, and yearly; and both have similar arrangements of entries that are listed alphabetically according to topic. At the start of these indexes, there are directions for use, a key to abbreviations, and a list of periodicals indexed. The following examples are taken from the Readers' Guide to Periodical Literature 1994 :

RADON DETECTORS
A Geiger counter on your face [eyeglasses react to radon; research by Robert Fleischer] il Discover v14 p 26 D '93
Radon monitor (I). P. Neher. il. Electronics Now v65 p 56-62+ Ja '94
Radon monitor (II. P. Neher. il. Electronics Now v65 p 66-70 F '94

RADON POLLUTION
See also
Radon detectors
Link of radon to lung cancer looks loopy [research by John S. Newberger] Science News v145 p 188 Mr 19 '94
Lung cancer risk in the home. L Katzenstein. il. American Health v13 p 39 S '94
New radon study : no smoking gun. R. Stone Science v263 p 465 Ja 28 '94
Radon revisited. N. Mead. il E : the Environmental Magazine v5 p 18-21 Ja/F '94
Radon : Dome concrete issues [study by Vern C. Rogers] Science News v146 p 191 S 17 '94

These entries start with the article title followed by the author's name, illustration included, title of magazine containing the article, and the magazine's volume number, the article's page numbers, and the magazine's date.

It is important to find the key word used to identify the topic you are researching. Articles about "radon" could also be listed under the topics of "gases," "pollution," "radiation," or "elements." Often a similar category will refer you to the correct category; such as, "See Radiation."

After determining a desired magazine, consult the library's periodicals holdings list to see if the library has the magazine in question. The periodicals holding lists are alphabetically arranged according to magazine titles and indicate the available issues. Either check the library floor plan to locate the current magazine section or ask the librarian about securing back issues. Many libraries allow you to access these periodicals online so you can avoid a trip to the library. In some cases, however, you may have to pay a fee to access articles online.

Visiting Working Examples

If you can see an actual example of your topic, do so and be prepared to take advantage of the visit. Places might include firms, shops, businesses, dealerships, hospitals, factories, construction sites, plants, mills, research facilities, public utilities, government agencies, retail stores, seminars, etc. When visiting, collect business literature and be prepared to interview representatives. It is wise to know as much as you can about your topic before visiting working example thus allowing you to talk in an intelligent manner and to ask intelligent questions regarding your topic.

Interviewing

If necessary, make an appointment. Introduce yourself and give your reasons for requesting the interview. Prepare your questions in advance and record the interviewee's responses to use as direct quotations in your technical research paper. This is a great way to get to know a prospective employer for future employment.

Writing, Phoning, or E-mailing Requests

When requesting research information through the mail, use the letter of inquiry format covered in Chapter XI. This takes more time than using e-mail, but it shows possible future employers that you have good writing skills and that you know how to write a letter.

When requesting research information over the telephone, conduct a phone interview. Before you make the phone call, make certain that you have a series of questions ready to ask. You want to sound knowledgeable when you speak to a company representative; so, make certain that you do your preliminary research before making the phone call.

When requesting research information using e-mail, remember that e-mail (electronic mail) is much faster than conventional mail, but you need to make certain that the e-mail is accurately written before you hit the "send" button.

Internet

When researching online, your evaluation process is complex. Make certain that you use reliable sources. Verify content from the Internet with other sources of information. Determine the credentials of the people you cite. Remember, the Internet is largely unchartered and unregulated territory. Wiki sites, in particular, may be changed by any user of the site. Sites ending in .edu are often safe sites in which to gain information. Remember that in your research paper you are responsible for the validity of your content. When using a library, most sources in a library are valid sources.

Format of the Research Paper's Eight Parts
(See example research paper that follows.)

Title Page

The title page is the first part of the technical research paper. The paper's title should be centered on the page both top to bottom and left to right. Capitalize the title's first and last words and all other words except articles, prepositions, and conjunctions. Use conventional punctuation. Follow the title suggestions made in Chapter II.

The student's name, the course title, and the paper's due date should be placed on three successive lines in the title page's lower, right-hand corner maintaining proper outside margins.

Table of Contents

The table of contents is a chronological listing of the paper's following sections: introduction, body, illustrations, conclusion, works cited, and glossary. Use the Roman numeral/capital letter outline format shown in Chapter IV. Only the body and illustrations are subdivided into capital letters.

Introduction

The introduction is the research paper's first part that is written in paragraph format and can contain endnoted information. The introduction includes four aspects of content:
1. Identification of topic, a complete sentence definition as covered in Chapter VI
2. Explanation of your personal reasons for choosing the topic
3. Review and evaluation of research methods
4. Summary of body's content, often in one sentence

Body

The body is the research paper's largest part, and it is divided into three to five main categories, the specific questions that you have researched. The body contains most of the endnoted information; especially quotations of experts, legal requirements, numerical statistics, specifications, and special hints or cautions. The body also contains references to illustrations when appropriate; such as, (See illustration 1, p. 10.).

Illustrations

The illustration section is a series of visuals (charts, diagrams, listings, blueprints, sketches, etc.) that bring your words to life. Most technical topics are impossible to understand without visuals. Neatly construct the visuals. Use color. Do not remove original illustrations from their sources to include with the paper. You may scan or cut/paste these visuals and import them into your paper. Remember to cite the sources of each visual so as to avoid plagiarism.

Give each illustration a title, document its source, and label its important parts.

Conclusion

The conclusion is the paper's last section that is written in paragraph format and that can contain endnoted material.

The conclusion's content is both a personal and general explanation of the topic's future. It includes your evaluation of the topic and its continuing research, development, or predictions.

Works Cited

Works cited (also called "bibliography") is an alphabetical list according to author's last names of all the sources that were used in writing the research paper. You do not have to cite a source in your paper to use this source for your works cited page. Oftentimes many of your sources will be used to develop common knowledge which is information which all sources agree upon and basically list verbatim.

Imagine that you are researching the topic of woodburning stoves and that you want to tell your readers that one pound of wood contains 7,000 Btu's (British thermal units) making cords of dense woods higher in Btu's than cords of less dense woods. If the 7,000 Btu's information were not cited as being determined by another person, your readers will assume that the 7,000 Btu figure is a calculation determined by you, the paper's writer. But how could you reach this conclusion? Do you have the time, materials, equipment, and expertise to test hundreds of species of wood? If your answer is "No," then tell your readers where you got the information. If you do not, you will be guilty of plagiarism, which is representing another's research or ideas as you own.

To show sources of borrowed information, you must show what was borrowed and where it was found. To show what was borrowed, you endnote it by putting the author's last name in parentheses after the borrowed content.

> "Since wood provides some 7,000 British thermal units (BTU) of heat per pound, it's easy to assume that heavier, more dense woods provide more heat per cord than lighter woods." (Self)

The word "Self" in parentheses is the author's last name, and the endnoted content is in quotation marks because the statement was copied word-for-word from its source.

To show where you found this statement by Mr. Self, you identify this one-author's book in the "Works Cited" section as follows:

Self, Charles. <u>Wood Heating Handbook</u>. Blue Ridge Summit, Pennsylvania : Tab

Books, 2010.

The endnoted content could also be paraphrased without quotation marks as follows :

A cord of wood provides approximately 7,000 Btu's (British thermal units) per pound; therefore, a cord of heavy, dense wood has more heat than a cord of lighter, less dense wood. (Self)

Since the previous example is not the author's exact words, quotation marks are omitted, but the author is endnote the same. The book's identification on the "Works Cited" page also stays the same. The alphabetical source identification includes the author's last name, the author's first name and possible middle name or initial, the title of the work, the place of publication (only the city is needed if the city is well known), the publishing company, and the copyright date. Notice the indentation, spacing, punctuation, and content in the following examples of books, magazines, and interviews. The MLA (Modern Language Association) Handbook for Writers of Research Papers gives more examples and detail, adding to these common publications.

Book by one author

Emerick, Robert Henderson. Underground Homes. New York : Mc Graw-Hill, 2000.

Print.

Book by two or more authors

Emery, John W., and Walter Walsh. How To Build and Furnish a Log Cabin. Boston :

Macmillan, 1999. Print.

Book, brochure, or pamphlet by a corporate author

General Motors Corporation. 1984 Chevrolet Corvette Shop Manual. Warren, Michigan:

GMC Service Publications, 2000. Print.

Reference book

"Kinetic Theory of Gases." Encyclopedia Americana. 2010 ed. Print.

Government publication

Montana. Department of Wildlife. Big Game Licensing and Hunting Procedures 2002.

Billings : Montana State Printing Office, 1996. Print.

Magazine article

Sellers, Michael and, William Alby. "A Classical Pool Complex." Architectural Digest

    May 2009: 192-197. Print.

Personal or telephone interview

Vickers, Susan A. Personal interview. 22 January 2012.

Internet site usage

(Visit www.mla.org for up-to-date format or check with your college or local library for the most current usage. This format changes almost yearly because of the ongoing changes on the Internet.)

Glossary

    The glossary is an alphabetical listing of the paper's technical terms and their definitions. The glossary gives only one of a heavily contextual term's meanings, the one that applies to the term's meaning in the paper.

    The glossary is very valuable to the novice reader because it saves him/her the time and effort needed to look up words in dictionaries and choose their appropriate definitions. The glossary does not slow the knowledgeable reader because he or she does not have to use it.

    The terms included in the glossary appear in the paper's introduction, body, or conclusion. These terms are not highlighted in the paper.

Eight-Step Procedure for Completing the Technical Research Paper

I. Choose a Topic and Write Questions To Research

II. Gather Information
    A. Visit the school library
        1. Check information in books
            a. In reference works
            b. In card catalog
        2. Review information in magazines
            a. From the Readers' Guide
            b. From the Applied Science and Technology Index
        3. Examine information in the Vertical File
    B. Visit other libraries for books, magazines, and unbound writings
    C. Visit working examples
    D. Conduct interviews
    E. Write or phone for information

III. Read Each Reference until Its Content Is Understood

IV. Make an Outline of the Body's Content and Choose the Paper's Title

V. Expand the Body's Outline into Paragraphs
    A. Group your research findings into the appropriate sections of the body's outline
    B. Join your research findings into paragraphs
    C. Refer to illustrations when appropriate
    D. Endnote borrowed information
        1. Quotations of experts
        2. Numerical statistics, specifications, costs, etc.
        3. Legal requirements
        4. Special hints or cautions

VI. Construct Illustrations

VII. Write the Introduction and Conclusion

VIII. Complete the Title Page, Works Cited, Glossary, and Table of Contents

Example of Technical Research Paper Preparation

Determining What To Research

Student's Major : carpentry
Specific topic : radon gas
Specific questions to research :
        1. What is radon gas?
        2. Why and how much radon is dangerous?
        3. How are radon levels tested?
        4. How are radon levels reduced?
        5. Can radon reduction measures be incorporated into new home construction?
        6. Are there building contractors who specialize in radon reduction? Are they licensed or specially trained?

Outline of Paper's Body

I. Testing Radon Levels
    A. Danger level
    B. Purchase of test kits
    C. Types of tests
        1. Short-term
        2. Long-term
    D. General testing procedures
    E. Professional Testing

II. Correcting Radon Problems
    A. Do-it-yourself options
    B. RCP certified contractor
    C. Reduction methods, according to foundation design
        1. Basement or slab-on-grade
            a. Sub-slab suction
            b. Drain tile suction
            c. Sump hole suction
            d. Block wall section
        2. Crawl space
            a. Active or passive ventilation
            b. Sub-membrane pressurization
        3. Basement, slab-on-grade, or crawl space
            a. Sealing
            b. Pressurizing
            c. Naturally ventilating
            d. Heat recovery ventilating

III. Preventing Radon during New Home Construction
    A. Cost savings
    B. Mike Nuess house example
        1. Location
        2. Awards won
        3. Operation of radon mitigation and heat recovery system
            a. Fresh-air entry
            b. Stale-air ventilation

[Example Technical Paper]

How To Rid Your Home of Radon

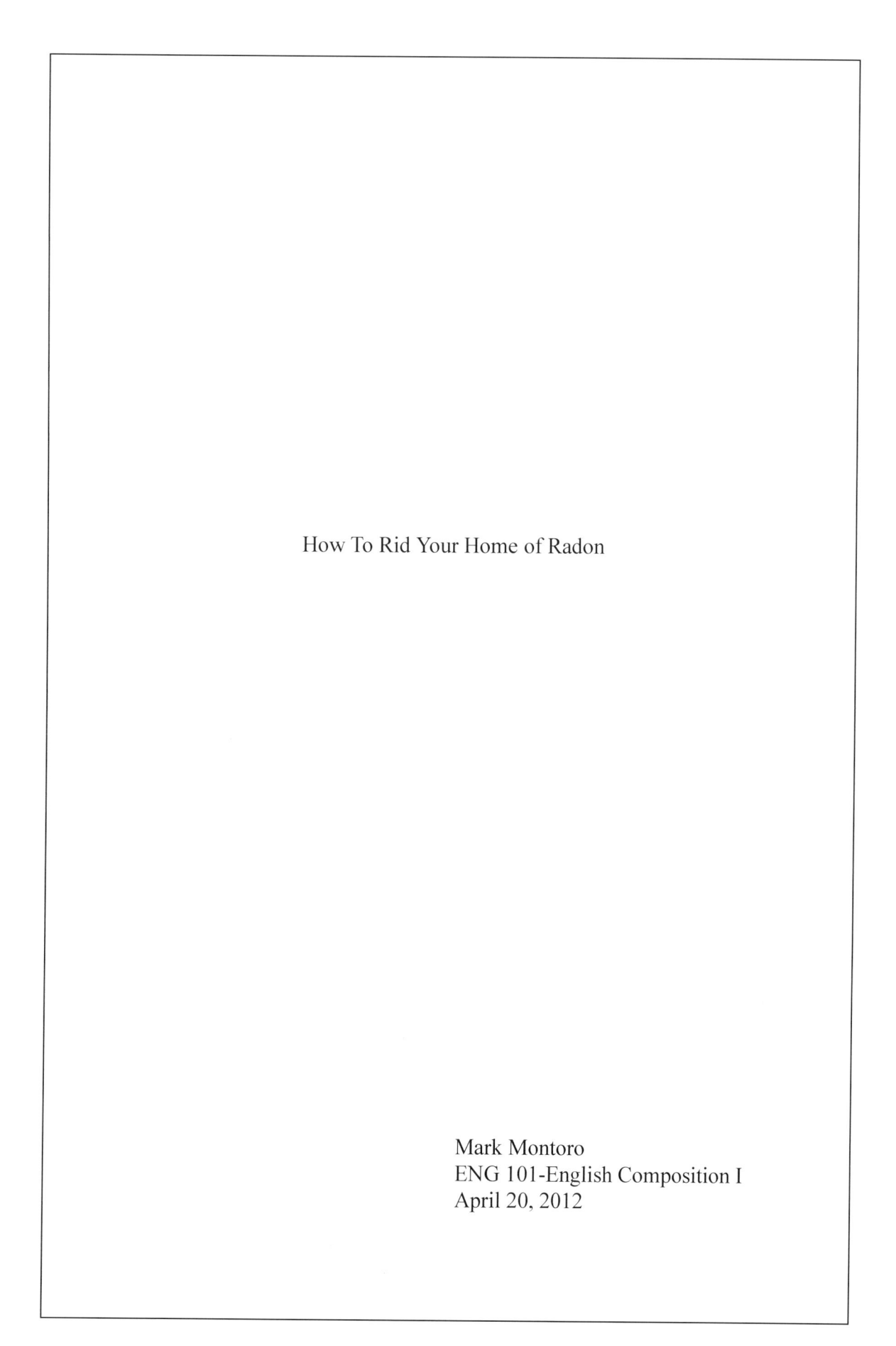

Mark Montoro
ENG 101-English Composition I
April 20, 2012

Table of Contents

I. Introduction 1

II. Body
    A. Identifying Radon Levels 3
    B. Correcting Radon Problems 4
    C. Preventing Radon during New Home Construction 6

III. Illustrations
    A. Radon Testing Firms in Lackawanna County of PA 7
    B. Radon Mitigation Firms in Lackawanna County of PA 7
    C. Typical Foundation Designs 7
    D. Radon Blocking Home of Michael Neuss 8

IV. Conclusion 9

V. Works Cited 10

VI. Glossary 11

Introduction

Radon (Rn) is a colorless, odorless, tasteless, inert radioactive gas that results from the decay of uranium, which is found in nearly all soils. "*Radon* is used loosely as the name of element 86 and all its isotopes, although the name emanation (symbol Em) is sometimes preferred for the element" ("Radon"). "Radon can be found all over the United States" (United States, "Citizen's Guide"). The radioactive gas travels from the ground through cracks and holes in a home's foundation and can build up to dangerous levels. Radon can also enter a home through well water, not public-utility water, but water entry is far less a risk than entry through air. "Nearly one out of 15 homes in the U.S. is estimated to have elevated radon levels" (United States, "Citizen's Guide"). "Pennsylvania has almost twice as many houses with dangerous radon levels as any other state" (United States, "Buyer's and Seller's Guide"). Any structure can contain radon, but our homes are our greatest source of radon because of their small size, air tightness, and the amount of time we spend in them.

Everyone who lives in a typical one or two-family home should know his or her home's radon level. "In fact, the Surgeon General has warned that radon is the second leading cause of lung cancer in the United States. Only smoking causes more lung cancer deaths. If you smoke and your home has high radon levels, your risk of lung cancer is especially high" (Pennsylvania, "Pennsylvania's Consumer's Guide").

Since I live in a ranch-style home in Pennsylvania with my parents, a younger brother, and a younger sister, I want to be sure that we are safe. Being a carpentry student, I believe that I should be familiar with building construction that prevents or corrects this life-threatening condition. I also would like to evaluate the feasibility of starting a business specializing in radon testing.

This paper's information was secured from our college's library, from an interview with a Pennsylvania licensed radon tester, and from United States EPA (Environmental Protection Agency) brochures, which were especially helpful. Articles that I found in various electronics magazines, often on how to build your own radon measurement devices, were too complicated to help me. The college library and our local public library had little radon information in their books. My most motivating piece of information came in conversation about my topic with a friend's wife who had her first child three months ago. She said that she was given a radon test kit by the hospital when she took her newborn home. I then asked my parents if our home was tested, and they answered, "No."

    This paper's body explains the identification, correction, and prevention of high radon gas levels in houses.

Body

Identifying Radon Levels

Radium decays by alpha emission to become radon gas as follows:

Mass units   226   222   4
Atomic Number: 88 Re→ 86 Rn   2 He
                         Radium   Radon   Alpha particle (Forney 88)

The amount of radon in the air is measured in pCi/L (picocuries per liter of air). "Fix your home if your radon level is confirmed to be 4 pCi/L or higher" (Pennsylvania, "Pennsylvania's Consumer's Guide"). Do-it-yourself radon measuring kits can be purchased through the mail, at hardware stores, or from general retail outlets. If your test results are to be used in home sale or radon mitigation, have the test conducted by a licensed, independent tester. Michael Stabinski, who is licensed by Pennsylvania and the federal government to perform radon testings, states: Ninety percent of independent testing is done as part of home inspection for sale. The licensed testing firms also perform before and after tests to confirm that radon mitigation has been done successfully (see fig. 1).

There are two ways to test for radon: the short-term test requiring from two to 90 days and the long-term test requiring more than 90 days. Charcoal canisters, alpha track, electret ion chamber, continuous monitors, and charcoal liquid scintillation detectors are most commonly used for short-term testing. Alpha track and electret detectors are commonly used for long-term testing. The long-term test gives a yearly average and is often used to confirm the evaluations of a short-term test that has indicated elevated radon levels.

All types of test kits contain specific directions for their use, and these rules generally apply to testing:

1. If doing a two or three-day test, close your windows and outside doors at least 12 hours before starting the test and during the test.

2. Do not conduct a two or three-day test during storms or unusually high winds.

3. Place the test kit into the home's lowest lived-in level in a frequently used room other than the kitchen.

4. Locate the test kit at least 20 inches above the floor away from drafts, high heat, high humidity, and exterior walls.

5. When the test is completed, send the test kit package to the specified laboratory for analysis. (United States, "Indoor Radon")

"The average indoor radon level is estimated to be about 1.3 pCi/l, and about 0.4 pCi/L of radon is normally found in the outside air" (United States, "Citizen's Guide"). Although all radon is a health risk, levels above 4 pCi/l should be corrected as previously quoted from EPA.

Correcting Radon Problems

If sealing obvious cracks and holes in basement floors and walls, the home owner is advised by the EPA to refer to the EPA's technical guide: *Radon Reduction Techniques for Detached Houses*. If the repairs are more technical, the state of Pennsylvania requires that you have a qualified contractor fix your home. A qualified contractor is one who has passed the EPA's Radon Contractor Proficiency Program (RCP). They carry a special RCP identification card and have their names recorded by your state's radon office for referral (see fig. 2). According to the EPA in 1992, "The average house costs about $1,200 for a contractor to fix, ranging from $500 to $2,500" (Pennsylvania, "Pennsylvania's Consumer's Guide"). A radon problem can be corrected before or after the gas enters a home although "the EPA recommends methods that prevent the entry of radon" (United States, "Buyer's and Seller's Guide").

The type of home construction affects the choice of radon reduction methods. Houses are classified according to three foundation designs as basement, slab-on-grade with concrete poured at ground level, or crawl space under the first floor (see fig. 3). Some homes have a combination of the three foundation designs requiring a combination of radon reduction techniques. For houses having a basement or a slab-on-grade foundation, radon can be reduced by using one of four techniques: subslab suction, drain tile suction, sump hole suction, or block-wall suction. The most common and most reliable radon reduction method is active subslab suction, also called "subslab depressurization." This method employs suction pipes inserted through the floor slab into the soil beneath. The number of pipes and their placement is determined by the ease of moving air below the slab and the concentrations of radon gas. A fan(s) connected to the pipes vacuums the radon gas from below the house and ventilates the gas into the outdoor air.

In homes having drain tiles to divert water from the foundation, suction can be used to pull radon from the tiles if they form a complete loop around the foundation. The sump hole suction system is a variation of the subslab and drain tile methods. This method caps the sump so that it can continue to drain unwanted water and also be the location for a radon suction pipe. If basements are constructed of hollow-block foundation walls, radon gas can be suctioned from the hollow blocks.

For houses having crawl spaces, the crawl space can be ventilated actively by using fans or passively by adding vents for natural air flow. Submembrane depressurization can also be used in crawl spaces. This method covers the ground with a heavy plastic sheet through which a vent pipe(s) and fan are used to draw the radon from below the plastic.

For homes with any type of foundation construction, radon levels can be reduced from sealing basement cracks and holes, by house pressurization using fans to blow air from upstairs or outdoors to the home's lower areas causing a pressure that prevents radon from entering, by natural ventilation from opening doors and windows, and by heat recovery ventilation (HRV, also called "air-to-air heat exchanger"), which uses the heated or cooled air being exhausted to warm or cool the incoming air. The previous four methods are only temporary or partial and should be used in combination with other methods.

Preventing Radon during New Home Construction

"New homes can be built with radon resistant features that minimize radon entry and allow easier fixing of radon problems that could occur later. These features cost less if installed during construction than if added to an existing home" (United States, "Buyer's and Seller's Guide"). Today's homes emphasize insulated tightness to reduce heat loss. Unfortunately, they also reduce radon loss. How to ventilate without losing heat is a problem.

One solution is the home of Mike Nuess of Spokane, Washington (see fig. 4). "Some of the homes in Spokane that were tested in a recent survey had radon levels 50 times greater than the U.S. EPA's recommended limit" (Stover 45). Mr. Nuess's home won the Energy Efficient Building Association's 2009 Design Competition Award and the EPA's 2009 Innovative Radon Mitigation Design Competition.

The two-story Neuss's house has one-foot thick walls (R-45) and triple-glazed windows, all creating an air-tight envelope. Radon is kept out by a pressurized crawl space below the first floor.

If radon does enter, it is mixed with the living areas' stale air being ventilated through the crawl space. Fans pull the home's stale air through ducts leading to the crawl space and traveling through a heat recovery system before emptying outside. Fresh air enters from a vent in the yard and travels through an underground tube, where the earth warms the air to at least 45° F before it enters a heat recovery ventilator and is ducted to the living areas. "The HRV provides about 45% of the required space heating, and the radon-mitigation and energy-conservation package added about $8,000 to the cost of the house," says Mr. Neuss (Stover 45).

Certified Radon Testing Firms in Lackawanna County of PA

Firm ID	Firm Name	Address	Phone
1713	Berardelli, Tom	312 N. Blakely St. Dunmore, PA	(717) 341-8844
1430	Homepro Resource Center	2040 Cedar Ave. Scranton, PA	(717) 343-5651
1570	Reality Inspection Service	312 N. Blakely St. Dunmore, PA	(717) 341-8844
1717	Stabinski, Michael E.	1016 Wood Land Way Clarks Summit, PA	(717) 586-5850
1662	Salima, Carol Anne	2042 Cedar Ave. Scranton, PA	(717) 343-5651

Fig. 1. This list represents certified local radon testing firms (Pennsylvania, "Radon Services Directory").

Certified Radon Mitigation Firms in Lackawanna County of PA

Firm ID	Firm Name	Address	Phone
1424	Salima, Joseph E.	2042 Cedar Ave. Scranton, PA	(717) 343-5651
1716	Vail, Jr., Robert J.	RR 2, Box 60 Jermyn, PA	(717) 254-2242

Fig. 2. This list represents certified local mitigation firms (Pennsylvania, "Radon Services Directory").

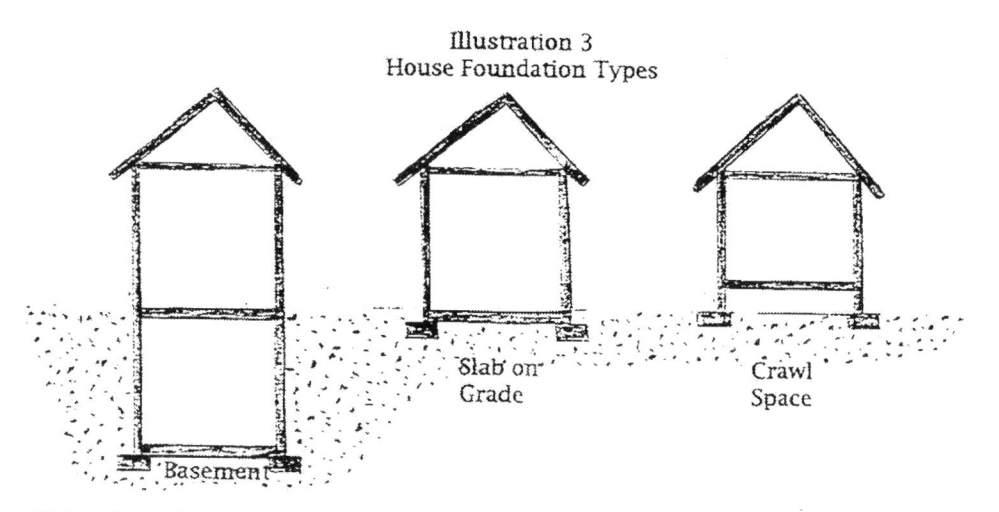

Illustration 3
House Foundation Types

Basement | Slab on Grade | Crawl Space

Taken from "How To Reduce Radon Levels in Your Home." September 1995

Fig. 3. This illustration shows three types of house foundations (Aberman 85).

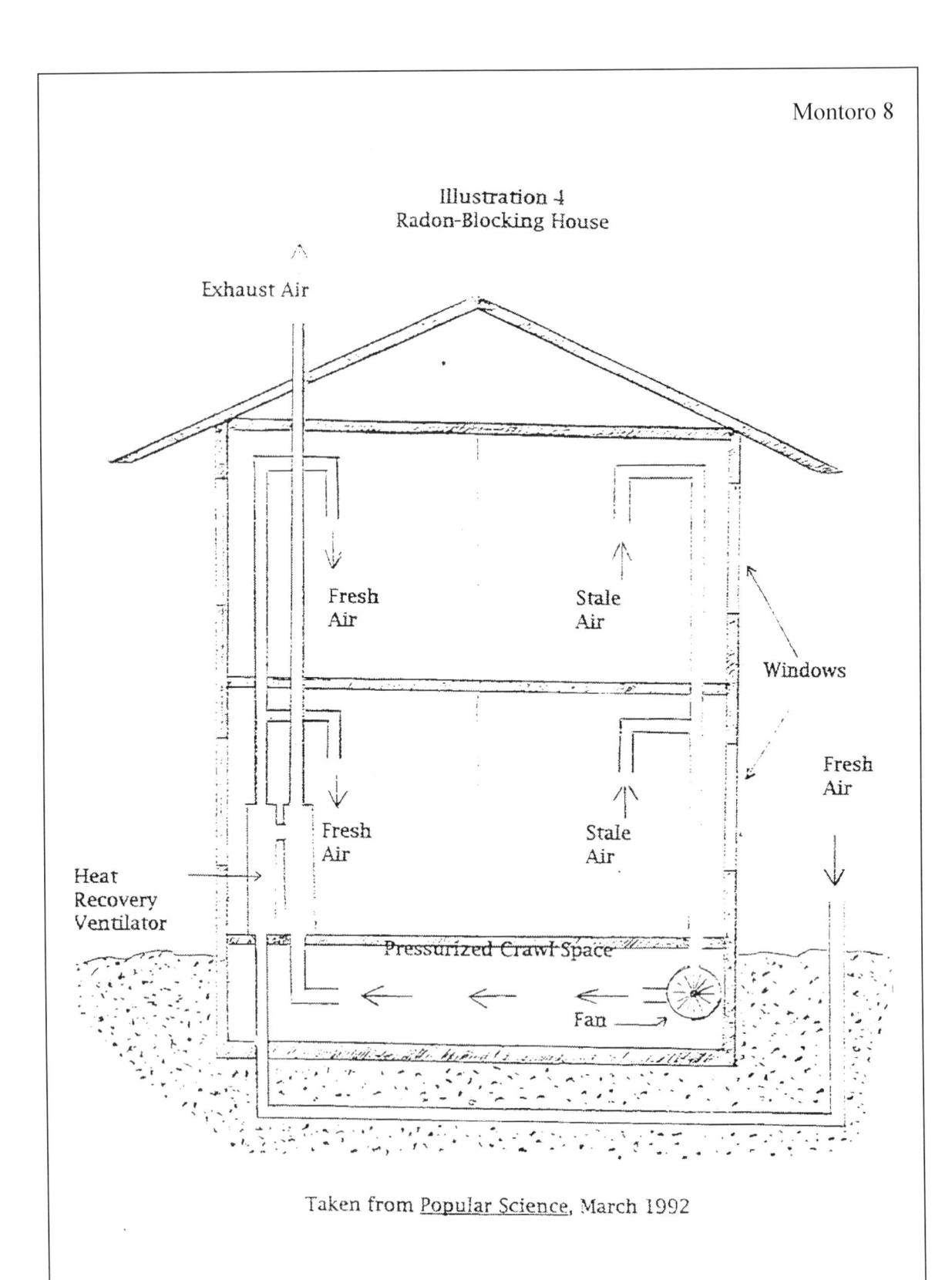

Fig. 4. This illustration depicts one new home's ways to keep radon out of the house (Stover 51).

Conclusion

Since radon gas will always be everywhere in varying degrees, we will all breathe it. The degree, however, is controllable in the most important area-our homes. Each time that we relocate or renovate our homes, we must be concerned with radon levels, not only for ourselves but also for our families.

I plan to buy a short-term radon test kit to measure the radon level in my family's home. If the short-term readings are above the recommended 4 pCi/L, I will conduct a long-term test. If the radon levels are still elevated, I will review the EPA's technical guide: *Radon Reduction Techniques for Detached Houses* to determine the corrections I can make and those that an RCP licensed contractor should make.

I also intend to research further the requirements of becoming a Pennsylvania state-certified tester and/or contractor. As I become more involved in the building construction career, I hope to use my knowledge of radon gas when designing or renovating.

While researching this topic, I became aware of recent studies questioning radon's cancer causing effects. The studies do not, however, contradict radon testing, correction, and prevention as covered in this paper. The studies have not yet been accepted by the U S Environmental Protection Agency.

Works Cited

Aberman, Richard. "How to Reduce Radon Levels in Your Home." *Home Improvement and Safety* 11.2 (2008): n. pag. Web. 7 Sept. 2011.

Forney, Lois W. *Chemical Principles for Life*. Englewood Cliffs, New Jersey: Prentice-Hall, 2006. Print.

Pennsylvania. Department of Environmental Protection. *Pennsylvania's Consumer's Guide to Radon Reduction*. Pennsylvania: Bureau of Radon Protection, 20 Mar. 2010. Web. 24 Oct. 2011.

Pennsylvania. Department of Environmental Protection. *Radon Services Directory*. Pennsylvania: Bureau of Radiation Protection, 2011. Print.

"Radon." *McGraw-Hill Encyclopedia of Science and Technology*. 7th ed. 2010. Print.

Strabinski, Michael E. Telephone interview. 9 Oct. 2011.

Stover, Dawn. "Radon-Blocking House." *Popular Science* 22.3 (2009): 45-53. Print.

United States. Environmental Protection Agency. *A Citizen's Guide to Radon*. Washington: GPO, 2010. Print.

United States. Environmental Protection Agency. *Home Buyer's and Seller's Guide to Radon*. Washington: GPO, 2011. Print.

United States. Environmental Protection Agency. *Indoor Radon and Radon Decay Product Measurement Device Protocols*. Washington: GPO, 2009. Print.

Glossary

active - requiring power to operate

alpha particle - two neutrons and two protons bound as a single particle that is emitted from the nucleus of certain radioactive isotopes in the process of decay

alpha track - a passive radon detector consisting of a small piece of plastic or film enclosed in a container with a filter-covered opening for excluding radon decay products

atomic number - the number of protons in an atomic nucleus

basement - the lowest habitable story of a building, usually below ground level

charcoal canister - a passive radon detector using activated charcoal to absorb radon for two to seven days as the absorbed radon undergoes radioactive decay

charcoal scintillation - a passive radon detector allowing the continual absorption and desorption of radon that is undergoing radioactive decay

continuous monitors - an active radon detector that samples the ambient air by filtering airborne particles as the air is drawn through a filter cartridge at a flow rate of about 0.1 to one liter per minute

contractor - one who does construction work as a business

crawl space - a low space below a floor for access to wiring or plumbing equipment

drain tiles - a four-inch pipe around a foundation to divert water from the foundation

ducts - enclosed passageways for a gas flow

electret ion chamber - a passive radon detector containing a charged electret (an electrostatically-charged disk of Teflon®) that collects ions formed in the chamber by radiation emitted from radon

element - a substance composed of atoms having an identical number of protons in each nucleus

emanation - an isotope of radon

foundation - the base upon which a building stands

heat recovery ventilator (HRV) - a device used to transfer heat from a vented air to an incoming air hollow block-a cinder block or concrete block with an empty center

inert - not readily reactive with other elements

isotopes - one of the two or more atoms having the same atomic number but different mass numbers

mitigation - the correction of an existing radiation condition

passive - not requiring power to operate

Pico Curie (pCi) - one trillionth of a Curie

Pico Curie per liter (pCi/l) - a unit of radioactivity corresponding to one decay every 27 seconds in a volume of one liter, or 0.037 decays per second in every liter of air

R-45 - a level of resistance to heat and cold used to measure insulation

radioactive - capable of the emission of radiation

radium - a luminescent, highly radioactive metallic element (symbol Ra) found in minute amounts in uranium ores

ranch-style house - a one-story dwelling, possibly having a basement

slab-on-grade - concrete poured on ground level

sump pump - a motorized suction device used to pull unwanted water from the lowest point in a basement

Surgeon General - the chief medical officer in the U. S. Public Health Service

triple-glazed window - three panes of glass separated by inert gas

uranium - an easily oxidized toxic metallic element (symbol U)

Exercises for the Technical Research Paper

Exercise 1 : Determining What To Research

>Directions : Choose one of the "Example Topics" from pages 25 and 26 and write a series of specific questions to research.

Exercise 2 : Determining What To Research

>Directions : Choose a topic for your technical research paper and write a series of questions to research.

Exercise 3 : Works Cited

>Directions : Arrange the following resources into the format of "Works Cited."

1. A magazine article titled "Recycling Plastic : An Award-Winning Idea" written by T. Adler and taken from page 181 of *Science News*, March 19, 1994, published by Science Service, Inc., Washington, DC

2. A book titled *How to Service and Repair Your Own Car* written by Richard Day and published by Harper and Row of New York City, New York, in 1973

3. A personal interview with Mr. James J. Stevens on September 20, 1995

4. A pamphlet titled *Removing Oil-Based Paint* written by the United States Environmental Protection Agency in 1996 and published by the Government Printing Office (GPO) in Washington, DC

5. An article titled "Marine Engineering" taken from the 1969 edition of the *Encyclopedia Americana*

6. A book titled *Fundamentals of Ecology* written by Eugene P. Odum, Walter D. Atkins, and Herman Martin, published by the W. B. Saunders Company of Philadelphia Pennsylvania, copyright in 1988

7. A magazine article titled "Venus Exposed" written by Robert Naeye and David Anderson, taken from pages 76-81 of *Discover*, December 1993

8. A magazine article titled "Vintage Radio" written by Marty Knight and taken from pages 63-66 of *Electronics Now*, January 1994 published by Gernsback Publications, Inc., of Farmingdale, New York

Exercise 4 : Technical Research Paper

Directions : Research your Exercise 2 topic and write a complete, eight-part technical research paper.

Chapter VI : Sentence Definition

People often ask what a thing is. Usually they are given only a partial, redundant, or structurally-incorrect definition. The best way to define is to use a sentence definition. Sentence definitions are used in many types of technical writing and speaking.

A sentence definition is a structurally-correct group of words (its verb is a form of "be") containing the term to be defined, its general class (group into which it falls), and its distinguishing characteristics (those aspects making it different from the other terms in its class). A sentence definition can be more than one sentence. Sometimes comparison (mention of similar qualities) and contrast (mention of different qualities) are helpful in longer sentence definitions.

The first sentence of the previous paragraph is a sentence definition of the phrase "sentence definition." The rest of the paragraph is also part of the longer sentence definition.

Examine this breakdown:

       term to be defined   "be" verb   general class
[A sentence definition] is [a structurally-correct group of words]
               distinguishing characteristics
[containing the term to be defined, its general class, and its

distinguishing characteristics.]

Compare to this shorter example:

[A wrench] is [a tool] [used to provide torque.]

Add the following if needed:

Wrenches can be classified as adjustable, combination, socket, allen, and torx. They all apply a twisting force on nuts, bolts, and some screws.

Three Errors To Avoid in Sentence Definition

Error #1: omitting any of the three parts

    Example: A car is for transporting people. (class omitted)
              A car is a motor vehicle. (distinguishing characteristics omitted)

    Correction: A car is a motor vehicle for transporting people.

Error #2: making a circular definition, using a form of the word being defined to define itself

Example: A technical research paper is a paper about a technical topic ("paper" and "technical" are used in class and in distinguishing characteristics.)

Correction: A technical research paper is a formal writing about a mechanical or scientific topic.

Error #3: using incorrect sentence structure (mechanics or syntax), especially the use of an adverb clause starting with "when" or "where" as a predicate noun

Example: An alloy is when (where) two or more metals are joined into one.

Correction: An alloy is a metal composed of two or more metals.

Note: Only one of the term's possible meanings is given in a sentence definition.

Correct Example: Sterilization is the hygienic process of killing germs.

Correct Example: Sterilization is the medical procedure of making a life form incapable of reproduction.

Exercises for Sentence Definition

Exercise 1 : Sentence Definition

Directions: Write sentence definitions of these. Use a dictionary if needed.

1. a hammer
2. a screwdriver
3. a welding rod
4. a set of saw horses
5. rabies
6. the word "technical"
7. a drill bit
8. exterior home siding
9. a job application
10. a die
11. a carburetor
12. a computer
13. emission control laws for cars
14. anatomy
15. a drill press
16. a rifle
17. a cellular telephone
18. scaffolding
19. plywood
20. a set of blueprints

Exercise 2 : Sentence Definition

Directions: Write sentence definitions for five or more terms used in your career choice.

Chapter VII : Summary

A summary is a shortening, condensation, digest, abridgment, abstract, or annotation of a writing's or speech's content. We summarize when a friend asks us how a movie was, what happened in school today, or what we did over the weekend. The movie lasted two hours, the school day seven hours, and the weekend 48 hours, but our summaries of them are much shorter.

Students and employees summarize when taking notes, reading chapters, writing reports and papers, giving speeches, and studying for tests.

Imagine that you are the employer of hundreds of employees of which 30 will be using a new, sophisticated machine whose operation is going to be explained in a five day, $500 seminar held 300 miles from your factory. Would you send all 30 employees for $15,000 and travel, lodging, food, and lost work time expenses? Or would you send one employee and ask him or her to summarize the five-day seminar into one day for the other 29?

When summarizing, you must understand the content to be summarized, the summary's intended audience, and the summary's purpose.

Understanding Content

Be alert, listen, take notes, read and reread, do research, and get experience. No one can accurately summarize what he or she does not know. When summarizing, document times, dates, pages, sponsors, writers or speakers, topics, and major points when documenting the summary's content. Summaries often omit details, examples, illustrations, and statistics.

The Intended Audience

If you were asked what you did over the weekend by your friend, mother, or boss, your weekend would be the same, but your summary might be different. You must determine what our audience knows, what it doesn't know, and what it needs to know.

The Purpose

There are three types of summaries: descriptive, informative, and evaluative.

The Descriptive Summary

It tells what a topic is about often in only one or two sentences. It is found in TV Guide listings like the following:

9PM 10 TOMORROW'S TECHNOLOGY (CC) 1:00 7736004 "Alternative Automobiles" reviews the most recent advances in electric and solar-powered vehicles.

11 WORLD WAR II (CC) 1:00 918036 "The Battle of Midway," fought between the US and Japan in June of 1942, was the most decisive naval battle of W.W.II. It was waged entirely between two aircraft carriers hundreds of miles apart.

12 JUNGLE STORIES (CC) 1:00 5320001 "Turf Battles" is the account of the ongoing struggles between lions and jackals for territorial rights. The lions' size and strength is balanced by the jackals' numbers.

in magazine tables of contents like the following:

12 **Pre-built Foundations**, by Jacob Graham, covers the design and placement of pre-built foundations.

20 **Home Plumbing**, by Ciecel Cedrick, explains the installation and repair of basement, kitchen, and bathroom plumbing.

35 **Heating Systems: Make the Right Choice**, by James P. Dow, details residential heating systems' pros and cons.

in abstracts, which are descriptive summaries of trade, technical, and clinical research and developments, like the following:

Ceramics in Automobiles is a 20 page prediction by Herman F. Appleton in Engineering Technology magazine, vol. 70, April 1996. The author concentrates on ceramic modifications to internal combustion engines, and he predicts their gradual introduction into production automobile engines.

in annotations, descriptive summaries of books, magazines, journals, reports, etc., like the following:

The Journal of Machine Trades is a monthly publication intended to keep machinists familiar with the most up-to-date advances in the industry. It includes regular articles on safety, testing, and quality control, as well as feature articles.

The Informative Summary

The informative summary gives the writing's or speech's main points objectively in approximately 10% of the original's size. The following is an informative summary of this chapter to this point.

A summary is a condensation of a writing or speech. Before summarizing, the writer must know his topic's content, its intended audience, and its purpose of describing, informing, or evaluating.

The Evaluative Summary

The evaluative summary is the same as the informative summary except that it adds the summarizer's subjective evaluation of the original's content, completeness, correctness, understandability, readability, and effectiveness. The following evaluative summary has its evaluative parts underlined.

The one day, six-hour plastic welding seminar held at Ohio State University on June 21 was <u>well attended, well planned, well equipped, and well taught.</u>

<u>Approximately 100</u> participants were divided into groups of 10 with one primary instructor for each group. The content for each group was the <u>same</u> although the order of presentation of information varied so that the example equipment could be used <u>efficiently and equally</u> by all participants. <u>Approximately 25%</u> of the seminar grouped all of its participants to hear the lectures of engineers.

The equipment was <u>state of the art</u> computers, robotics, measuring instruments, testing devices, and test materials. Besides the theory of the preceding, participants were <u>often</u> given the chance to get hands-on experience with the equipment.

<u>Most</u> of the instructors had teaching experience on the college level. All of the engineers were employed by local industries.

<u>I think that this seminar was excellent but that there should have been some input from the manufacturers of the equipment demonstrated.</u>

Exercises for Summary

Exercise 1 : The Descriptive Summary

Directions: Write descriptive summaries of as many of these topics as possible:

1. Each of the subjects you are taking this semester in school
2. A chapter of the textbook in your major study
3. A guest speaker's presentation that has occurred in one of your classes
4. A television show or series that you enjoy
5. A book that you've read

Exercise 2 : The Informative Summary

Directions: Write informative summaries of as many of these topics as possible:

1. The three types of summary
2. Any chapter of this textbook other than this one
3. A job that you have had
4. Your favorite sport's main rules
5. A project that you've worked on

Exercise 3 : The Evaluative Summary

Directions: Write evaluative summaries of as many of these topics as possible:

1. The most interesting project you have ever been involved in
2. The worst job you ever had
3. Your favorite magazine
4. A seminar, exhibit, or demonstration you have attended
5. An organization of which you have been a member

Chapter VIII : Instructions

Instructions tell how to do something like how to drive to a site, how to apply for a job, how to write a business letter, how to assemble a gas grill, how to troubleshoot an electrical system, or how to repair a bicycle's flat tire. Instructions are given for travel and operation in a numbered chronological order, and the subjects of the steps are usually "You" understood.

Travel Instructions

When telling someone how to get somewhere by vehicle or foot, we usually start with the first direction followed by the second, third, etc. Avoid this impulse and introduce these directions with an explanation of starting point, destination, direction, distance, and travel time. Illustrations are often helpful.

Driving Directions to Penn State University from Scranton, Pennsylvania

Penn State's main campus in State College is located in the exact center of Pennsylvania, approximately 150 miles south/west of Scranton, and can be driven in three hours by following these directions:

1. Take Interstate 81 south to Interstate 80 West (approximately 45 miles).
2. Take Interstate 80 west to the Bellefonte Exit 24 (approximately 90 miles).
3. Take Pennsylvania Route 26 west into State College (approximately 15 miles).

Note: Details like diagrams, exact starting point in Scranton, directions from start to Interstate 81 south, and land-marks passed were omitted from the previous instructions. Sometimes they are very important depending upon the complexity of the directions.

Walking Directions from Home to Work

The Phillip's Steel Works, Inc., Athens, Texas, is two and one-half miles north of my home at 563 Smith Street. The distance can be walked in 40 minutes by using these directions around the Athens Reservoir #3:

1. Follow Smith Street north for two blocks to School Road.
2. Stay on School Road east for one block.
3. Take short cut through Athens High School's outdoor practice fields by crossing the softball field, practice football field, and soccer field to arrive at the north end of the Interfaith Cemetery on Wallace Avenue.
4. Follow Wallace Avenue north, past the cemetery and two more blocks to the railroad tracks.
5. Follow the tracks north/west for one mile.
6. Enter the rear gates to loading docks at Phillip's Steel.

Diagram

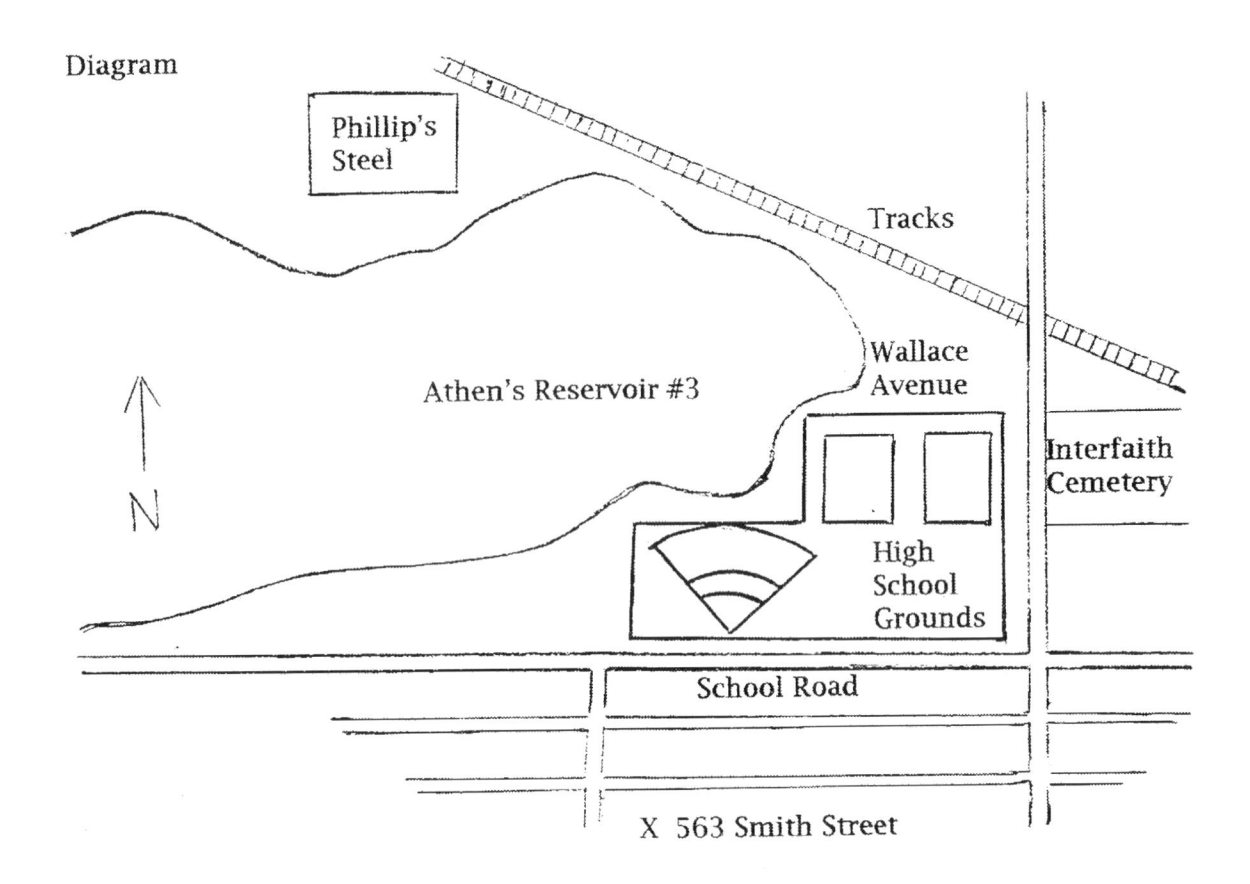

Operational Instructions

When telling someone how to design, fabricate, assemble, repair, service, install, or build, we are giving operational instructions. Like travel instructions, operation instructions are given chronologically, use a numbered list, have the subject "You" understood, and sometimes require details, especially if there are cautions, hints, or explanations of step completion needed. The use of detail depends upon complexity, audience, and purpose.

How to Change an Automobile Tire

Complexity : Simple
Audience : One who has changed many auto tires
Purpose : A wheel and tire replacement manual that might be found inside a car's trunk

   1. Secure car safely.
   2. Prepare materials and equipment.
   3. Loosen lug nuts.
   4. Jack car up.
   5. Remove lug nuts and tire.
   6. Replace second tire and hand tighten lug nuts.
   7. Lower car.
   8. Tighten lug nuts and replace materials and equipment.

Complexity : Same but more complex for a new driver
Audience : A driver-training student
Purpose : To explain thoroughly how to change an automobile wheel and its tire under various conditions

1. Secure auto safely.
   a. Park out of traffic on solid, level ground.
   b. Turn engine off and place automatic transmission in "Park." If possible, place standard transmission into low gear if front is to be jacked up but reverse gear if the back is to be jacked up.
   c. Apply emergency brake and, if near traffic, emergency flashers.
   d. Remove possible passengers.
   e. Block wheels. Wedge brick-sized sturdy material between the grounded wheels and the direction the car would likely roll.
2. Prepare materials and equipment by getting new tire and jacking devices from their holding areas.
3. Loosen lug nuts (the five hex nuts that hold the wheel to the axle). These might be covered by a hub cap, which is usually removed by pinching it from the wheel with the flattened end of the tire iron. (See Illustrations 1, 2, and 3). **Caution:** The nuts are only "broken," freed so they can be easily turned off later.

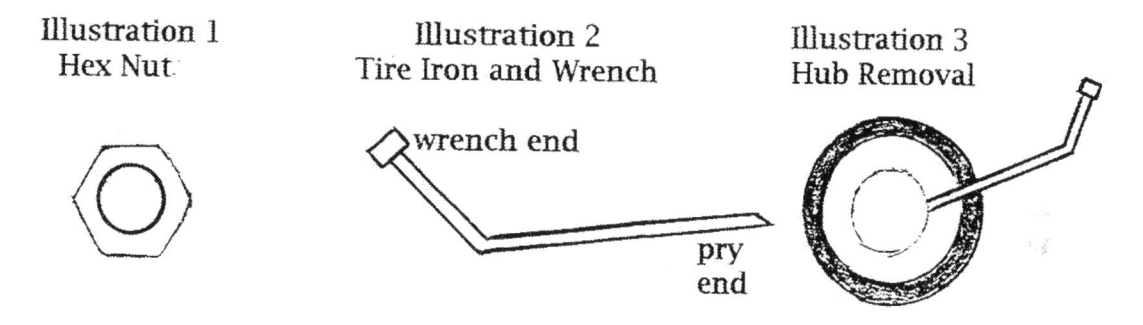

Illustration 1
Hex Nut

Illustration 2
Tire Iron and Wrench

wrench end

pry end

Illustration 3
Hub Removal

4. Raise car up enough so that the new (inflated) tire will clear ground. Use jacking or twisting motion depending upon type of jack. See Illustrations 4 and 5 for operation of lifting devices, and see Illustration 6 and 7 for placements of lift.

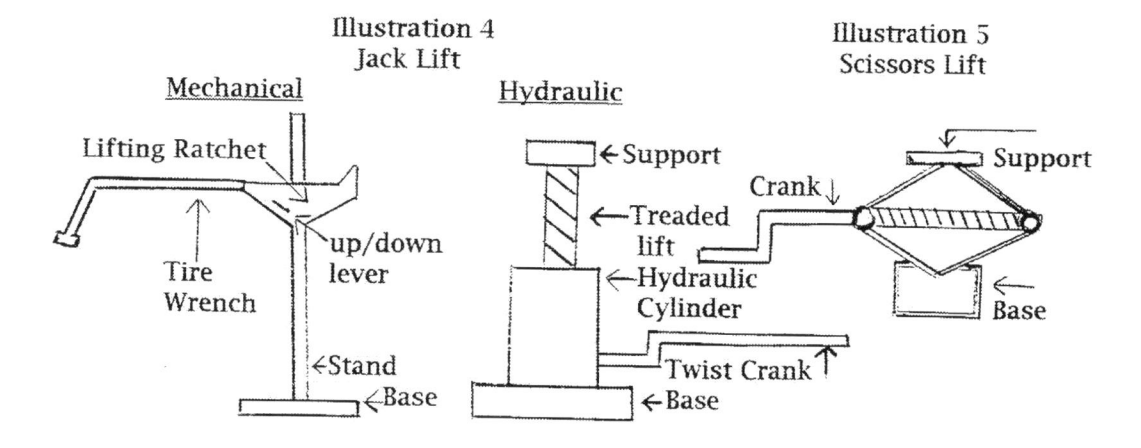

Illustration 4
Jack Lift

Mechanical

Lifting Ratchet
Tire Wrench
up/down lever
←Stand
←Base

Hydraulic

←Support
←Treaded lift
←Hydraulic Cylinder
Twist Crank↑
←Base

Illustration 5
Scissors Lift

Crank↓
Support
←Base

Illustration 6
Side Placement
(under frame)

Illustration 7
Front or Rear Placement
(under bumper)

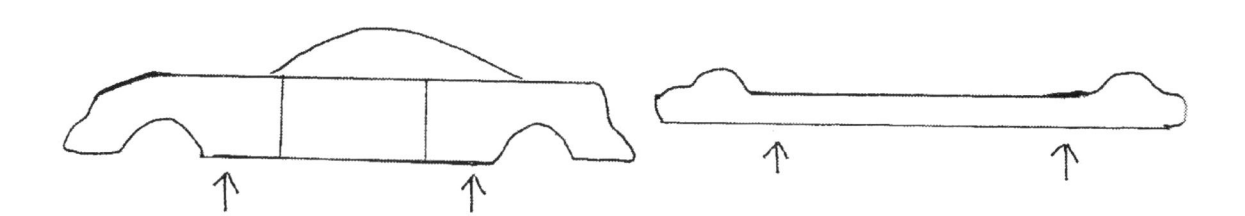

5. Remove lug nuts and wheel and tire
   Caution: Be careful not to jar the vehicle when removing the wheel and tire. Grasp the tire at right and left centers.
6. Replace second tire and hand tighten lug nuts.
   a. Be careful not to jar the vehicle.
   b. Tighten the nuts only until they snug the wheel to the axle.
      Caution: Complete tightening could dislodge the car from the lift.
7. Lower car. Reverse the jacking or twisting movements of raising the car.
8. Tighten lug nuts. Apply considerable force to be sure that the wheel is tight and safe.
9. Replace materials and equipment.
   a. If necessary, tap hub cap onto wheel.
   b. Store original tire and lifting equipment.

Exercises for Instructions

Exercise 1 : Travel Instructions

Directions: Write travel directions from your school to your home.

Exercise 2 : Travel Instructions

Directions: Write travel directions for one or more of the following:
1. From your city to another city by car
2. From your home to a friend's home on foot
3. From a parking area to a location in the woods; such as, a fishing area, deer stand, camping area, etc., on foot or by vehicle
4. From a room or section of a building, factory, school, etc., to another area of the structure on foot
5. From your state to another state by airplane

Exercise 3 : Operational Instructions

Directions: Write operational instructions to an inquiring customer for one or more of the following :
1. How to change a car's antifreeze
2. How to plug a tubeless tire of a car
3. How to install electrical wall receptacles
4. How to prepare for the birth of a pet
5. How to troubleshoot for a problem related to your technology major
6. How to complete a construction project
7. How to clean and service a machine
8. How to assemble a gas grill, lawn mower, piece of exercise equipment, etc.
9. How to operate a recreational vehicle
10. How to install a manual carburetor choke, a window, a door, sidewalk, etc.

Exercise 4 : Operational Instructions

Directions: Choose a topic and write two sets of operational instructions one for a beginner and one for an expert.

Chapter IX : Description of a Device

A device is a mechanism with moving parts; such as scissors, an adjustable wrench, a Jacuzzi®, a gasoline engine, or an automobile. Describe a device by explaining its purpose, its overall appearance, and its parts and their interaction. A device can be described at rest and/or in motion. There are two types of device descriptions: general and specific.

The Purpose

The purpose of a device is what it does; possibly including, when, where, and how, as well as by whom. Often, a sentence definition (Chapter VI) can include all of the purpose. For example: A radio is an electronic device used to receive audible signals of communication encoded in electromagnetic air waves and transmitted by radio broadcasters. A two-way radio can receive and transmit signals.

The Overall Appearance

What does the device look like externally? Give its size, shape, weight, material, color, and finish. Comparison to a more well-known item is often helpful. For example: Pipe wrenches are made in various sizes but always of metal. At the end of a long handle for leverage, there are two serrated jaws, the upper one being adjustable. A pipe wrench looks like a printed capital letter "F."

Its Parts and Their Interaction

Describe each part of a device by giving its purpose, its overall appearance (size, shape, weight, etc.), its location, and its interaction with other parts. A spatial order (Chapter IV) is best for the description of a device at rest, and chronological order is best for the description in motion. Illustrations are also helpful. For example: The yoke and spider universal joint allows parts of a machine not in line some movement in any direction while transmitting rotary motion. The "spider" is a pair of crossed and joined steel rods that are free to swivel in the opposing yokes they connect.

Yoke and Spider Universal Joint

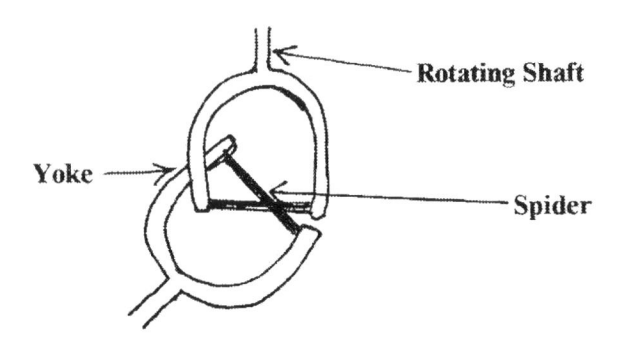

General Description

The general description treats the device as though it is representative of all devices in its group. It stresses the purpose, overall appearance, and parts and their interactions that are usually associated with the device. Examples are a ball-point pen, a flashlight, a portable circular saw, or a motorcycle. This description is used in an encyclopedia.

Specific Description

The specific description focuses on a particular brand, model, or style of device. The description concentrates on features making the device different from others in the same group; such as, size, ease of use, performance, guarantee, accessories, price, etc. The device's purpose, overall appearance, and parts and their interactions are usually not listed unless they describe the device's uniqueness. Examples are a Bic® ball-point pen, an Eveready® D-size flashlight, a Skill® portable saw (model 75E), or a 1996 Harley Davidson®Buell®motorcycle. This description is used in advertising.

Example General Description
of an Electric Doorbell in Operation

An electric doorbell is an attention-getting device producing an audible ring (similar to a telephone's) when the circuit is completed at the button by the person wishing admittance.

The doorbell mechanism, protected in a metal or plastic box approximately 5" X 5" X 2", includes a bell, an armature with its clapper and spring, an electro-magnet, and a contact screw. (See illustration 1.) The completed system also requires a power source, wiring, and a press-button switch. (See illustration 2.)

When the doorbell button is pushed, the circuit is completed allowing electricity to flow from its source (a dry-cell battery or a transformer) to the electromagnet. The current activates the electromagnet that draws the pivoted armature to the magnet and armature's clapper to the bell. However, when the armature is drawn to the magnet, the circuit connection between the armature spring and the contact screw is broken, and the magnet ceases to operate allowing the armature to return to its original position. As long as the doorbell button is held in, the process repeats the striking of the clapper against the bell, which produces a continuous ring.

The electric doorbell principle is also used in telephones, automobile horns, school bells, fire alarms, and alarm clocks.

Illustration 1

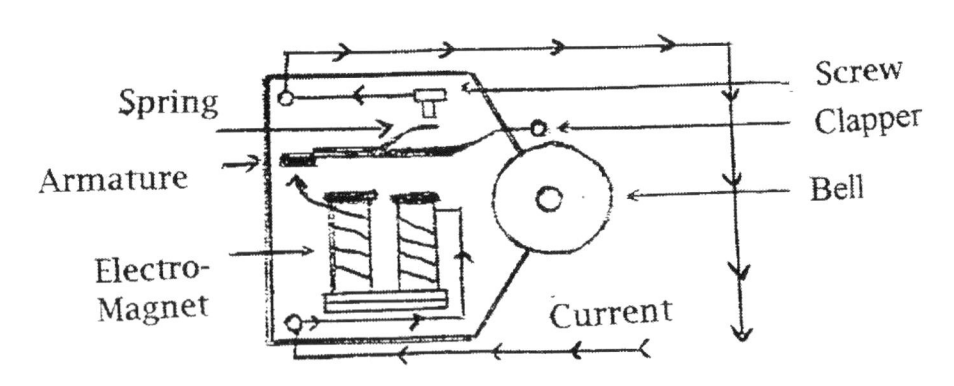

Illustration 2

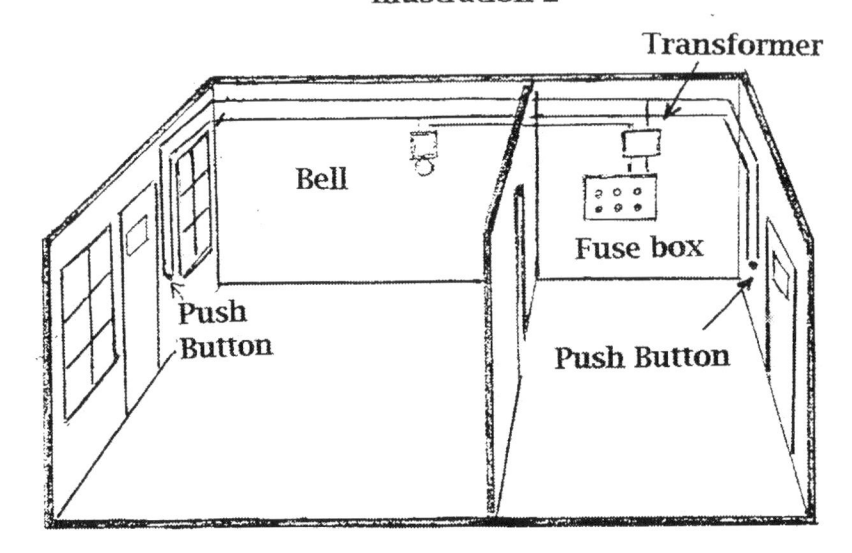

Example Specific Description of the Decorative
Electronics® Model-1010 Electric Doorbell

The M-1010 electric doorbell manufactured by Decorative Electronics is tasteful and versatile because of its quality construction and choice of sounds.

The doorbell button is lighted and enclosed in a sturdy brass frame measuring 2 1/4" X 2 1/2". The bottom of the frame holds a brass identification insert on which up to 15 letters of a name or address can be engraved.

The doorbell mechanism is contained in a 7" X 6" X 3" metal box that is front, flush mounted to a wall where it is hidden behind a brass, ventilated cover. At the cover's base, there is a slide switch that allows the choice of three doorbell sounds: standard buzzer, classical chimes, and wind chimes. The kit includes all mounting hardware, transformer, and 200 feet of wire.

This doorbell sells for $69.95 postage paid from Decorative Electronics, Co., 2706 Laurens Road, Greer, South Carolina 29651, phone (864) 288-2621.

Brass Doorbell Button   **Brass Cover**

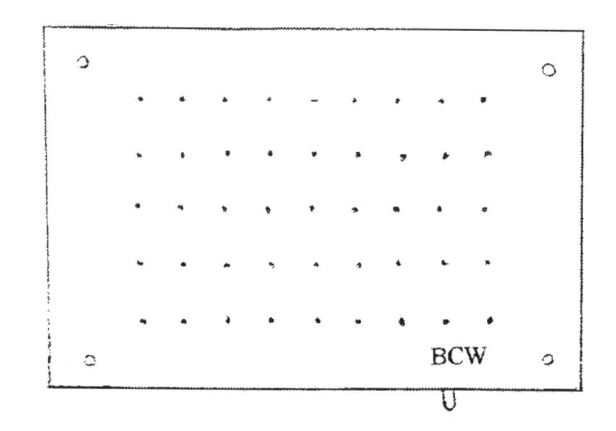

Exercises of Description of a Device

Exercise 1 : General Description of a Device

Directions: Write a general description of one or more of the following:
1. a tread mill
2. a roll-top desk
3. a paint roller
4. a nut and bolt
5. a flashlight
6. an electric end-table lamp
7. a master cylinder of a hydraulic brake system
8. a manual drill
9. a hydraulic jack
10. a block and tackle (winch)

Exercise 2 : Specific Description of a Device

Directions: Arrange these details into a description of a specific device. Imperial Ceiling Fans are available in either 36," 42," or 48" blade diameters. Both five and six blade arrangements are offered. The blades are made of unstained oak, walnut, or cherry. The metal exterior has a solid brass decorative ring, motor cover, and blade attachments. The three-speed motor has the highest UL ratings for energy efficiency and usage hours. The motor operates in both directions. Most standard light fixtures can be attached. All of the above combinations are possible; such as a 48"diameter, five blade, and cherry model. The prices are five blade, 36" diameter for $100; five blade, 42" diameter for $110; and five blade, 48" diameter for

$120; six blade, 36" diameter for $120; six blade, 42"diameter for $130; and six blade, 48" diameter for $140. All wood choices are of equal cost. The shipping charge per unit is $10 anywhere in the continental US from Imperial Ceiling Fans, 414 Seventh Avenue, New Holland, Pennsylvania 17557. Phone (717) 354-4311. Pennsylvania residents add .06% sales tax.

Exercise 3 : General and Specific Description of Devices

Directions: Choose a specific model, brand, or style of device used in your major study and write both a general description and a specific description of it.

Chapter X : Description of a Process

A process is a procedure that can not be done by one person. The purpose in describing a process is to explain what is done by nature or by men often using machinery (a work process). Natural processes are the forming of hail, diamonds, coal, petroleum, earthquakes, or mountains; the changing of day to night, of new moon to full moon, or from fall to winter; and the maturing of people, animals, plants, or other life forms. Work processes done by men using machinery are the mining of diamonds, coal, or iron ore; the building of seismographs, ore drilling rigs, electrocardial graph machines, or the Hubbell Telescope®; the operating of a factory, airport, or government; and the performing of a medical operation, maintenance on a city's streets, protection of a country.

When giving a description of a process, use a gerund (a verbal ending in "ing" and being used as a noun) for the title of the process and for each of the process's major steps. The steps of a process are often given in chronological order. The verbs used in the description of a process are often in their passive voice when the subject receives the verb's action rather than performs the verb's action as active voice verbs do. A passive voice verb is always accompanied by a form of the helping verb "be" (<u>is</u>, <u>are</u>, <u>was</u>, <u>were</u>, <u>am</u>, <u>be</u>, <u>been</u>, <u>being</u>).

Passive voice verbs emphasize that the radar is not expected to perform the process. In the previous sentence "emphasize" is an active voice verb, and "is expected" is a passive voice verb. If describing the starting of a nuclear power plant, in active voice we might write, "Ten technicians start the turbines." In passive voice we might write, "The turbines are started by ten technicians."

Check these chronological listings of gerund phrases for the major steps in the natural process of the rusting of metal and the group process of hiring an employee. The gerund phrases are expanded into the completed process descriptions that follow the major steps.

Rusting of Metal

1. Exposing
2. Ionizing
3. Deteriorating

Hiring an Employee

1. Writing job description
2. Advertising the job opening
3. Accepting résumés
4. Having applications completed
5. Interviewing applicants
6. Checking references and transcripts
7. Administering physical examination and possible skill tests
8. Offering the position
9. Accepting the hiring contract

Expanding Major Steps of a Process

Rusting of Metal

The rusting of metal causes billions of dollars of damage each year. A rusting car is said to have "cancer" because as cancer eats the human body, rust eats the car's body. Rusting is an electrochemical process in which there is a flow of current between an anode and a cathode, causing the anode to be eaten away. There are three steps in rusting: exposing, ionizing, and deteriorating.

Exposing the metal to atmospheric conditions, mainly oxygen and moisture, begins the reaction. Paint, oil, and grease are excellent shields against the atmosphere, but when these shields are removed, the metal is exposed.

After the metal is exposed, it starts ionizing. Ionizing is the forming of micro-galvanic cells or batteries, consisting of an anode (positive terminal) and a cathode (negative terminal). Electrons flow between these terminals.

The third stage is the deteriorating of the anode from the continuous current flow while the cathode remains intact. For example, if aluminum screws are used in iron, the iron (cathode) would not rust. However, the screws (anode) would corrode much faster than they would were they not contacting the iron.

Once started, rusting does not stop until the rusting metal deteriorates into the same oxides found in the original ores.

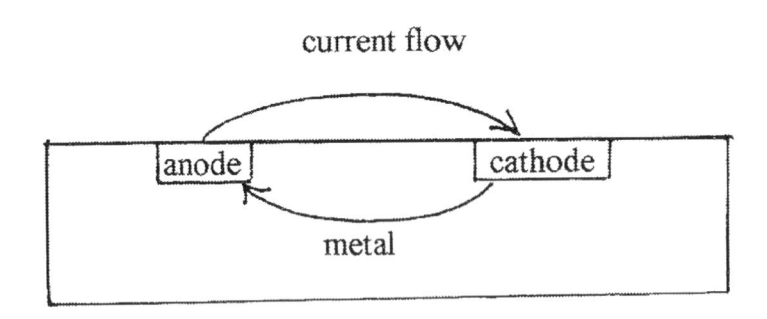

current flow

| anode | cathode |

metal

Hiring an Employee

When an employee is hired by a large company, many people are required to complete the nine stages of hiring.

The first step is writing the job description to identify the conditions of employment and the employee's requirements from skill to pay. Management personnel are generally responsible for this, as well as for the second step of advertising the opening through newspapers, agencies, company bulletin boards, or schools.

The third, fourth, fifth, and sixth steps are usually carried out by the personnel manager, who is responsible for accepting résumés, having applications completed, interviewing applicants, and checking references and transcripts. The applicant's qualifications, as shown in steps 3-6, are also evaluated by the personnel manager.

The seventh step is administering a physical examination, which is done by a doctor and his or her assistants. Other testing of skills may be administered by specialized personnel.

The eighth step, offering the position, and the ninth step, accepting the hiring contract, are done by employment managers, supervisors, owners, boards of directors, etc.

Exercises for Description of a Process

Exercise 1 : Passive Voice Verbs

> Directions: Make a list of the passive voice verbs used in "Hiring an Employee" on the previous page.

Exercise 2 : Passive Voice Verbs

> Directions: Make a list of passive voice verbs used in "Rusting of Metal" on the previous page.

Exercise 3 : Description of a Natural Process

> Directions: Make a chronological list of gerunds to identify the major stages in as many as possible of the following natural processes:
>
> 1. The Forming of Radon Gas, Lightening, Clouds, Hurricanes, etc.
> 2. The Drying of Concrete
> 3. The Digesting of Food
> 4. The Changing from Caterpillar to Butterfly or from Tadpole to Frog
> 5. The Growing from Puppy to Adult Dog, from Kitten to Adult Cat, fro Calf to Adult Cow, etc.
> 6. The Developing from Seed to Flower, Weed, Bush, or Tree
> 7. The Decaying of Teeth
> 8. The Petrifying of Wood
> 9. The Fermenting of Fruit
> 10. The Developing of a Disease's Stages (AIDS, Hoof and Mouth Disease, Rabies, etc.)

Exercise 4 : Description of a Natural Process

> Directions: Develop the stages for one of the Exercise 3 topics into a complete descriptive writing.

Exercise 5 : Description of a Work Process

Directions: Make a chronological list of gerunds to identify the major stages in as many as possible of the following processes done by men with machines.

1. Recycling Aluminum, Glass, Plastic, etc.
2. Building a Mall of Retail Stores
3. Mining for Iron Ore, Coal, Diamonds, etc.
4. Smelting of Iron, Zinc, Gold, etc.
5. Caring for a City
6. Paving a Highway
7. Extinguishing a Forest Fire
8. Transplanting an Organ
9. Advertising a Product
10. Manufacturing a Product (in a place you worked)

Exercise 6 : Description of a Work Process

Directions: Develop the stages for one of the Exercise 5 topics into a complete descriptive writing.

Chapter XI : E-Mail/Business Letters

Much business is done partially or totally by e-mail, mail or fax. E-mail is nothing more than electronic mail. It is a much faster process of communicating with others in the world. Remember that despite it being sent to the recipient by electronic means, the contents of each e-mail should be similar to that of a business letter. E-mail may be formal or informal. Typically, we send informal e-mails to friends, etc. In the business world we must send formal correspondence at least until we get to know one another.

The business letter is formal in format and content, having at least six parts. There are a number of styles (full-block, modified-block, and indented), but we will emphasize the full-block format in which all parts start at the left margin.

The contents of a business letter's body depends upon the letter's purpose: to inquire, to order, to claim, to adjust, to sell, or to apply for a job. Since the previous transactions often take more than one letter to complete, keep the letters you receive and copy those you send.

Appearance

I. Use a good quality white, unlined, 8 1/2" X 11" business-letter stationery and a No. 10 size (4 1/8" X 9 1/2") envelope. Letterhead stationery used by firms have preprinted headings needing only the letter's date. Type if possible.

II. Leave at least an inch margin around the page. Letters with short bodies should be centered, having larger than one inch margins at top and bottom.

III. Give the letter's purpose in the first sentence of the letter's body. Paragraphs are short, usually not more than eight lines. Do not abbreviate (except for "Mr.," "Mrs.," "Ms.," "Dr.," "Jr.," "Sr.," etc.), do not use contractions ("it is" vs. "it's," etc.), and do not hyphenate words ending lines to continue them on the next line. Put the whole word on the next line. Never start the letter with "I am writing this letter to you because . . ." because it is obvious. Start with the part following "because."

IV. Double space between the parts of a letter and between paragraphs. Single space within parts and within the paragraphs.

V. Always include the business letter's six required parts: heading, inside address, salutation (greeting), body, closing, and writer's signature.

The following is an example of the six required parts in full-block format:

Six Required Parts in Full-Block Format

1139 Stafford Avenue [The heading doesn't have the writer's
Hickory, North Carolina 28602 name. Notice the commas between city
August 23, 2012 and state and between day and year.]

<u>Popular Science</u> [The inside address spells, abbreviates, and
Mr. Earl Williams, Editor punctuates a company name as the company
2 Park Avenue does. The addressee's name is prefaced with
New York, New York 10016 a title of respect; such as "Mr.," "Mrs.," "Ms.,"
 or "Miss."]

Dear Mr. Williams: [The salutation is "Dear" followed by a title of
 respect, the addressee's surname, and colon.]

In the full-block style, paragraphs are not indented.

When possible, stress the reader before the writer. Use "you," "your," and "yours" vs. "I," "me," "my," and "mine." Be positive by telling what is rather than what is not. Always be courteous.

The last paragraph is often a one sentence statement of appreciation.

Very sincerely yours, [The closing is "Sincerely," "Sincerely yours,"
 "Cordially," "Respectfully," etc. Capitalize
 only the first word and end with a comma.]

James J. Johns [The signature is cursive and then typed or
 printed below the cursive.]

VI. On the second, third, etc., pages of a business letter (even those using letterhead stationery for the first page), give the addressee's name, the letter's date, and the letter's page number. In the full-block format, this information is given at the top left margin, on three lines, and in the above sequence. The second or last page should contain at least two lines of the letter's body, the closing, the signature, etc.

The following is a second-page format example:

Mr. Earl Williams
April 23, 2012
Page 2

Business letters seldom take more than one page. By including at least two lines of the letter's body, the writer is more secure in his or her not having the preceding pages changed and his or her closing, signature, etc., attached.

Sincerely yours,
James J. Johns
James J. Johns

VII. If needed, include one or more of these four special parts: attention line, enclosure, typist's initials, and copy notation.

A. The attention line helps in mail delivery when the writer does not know the addressee's surname. The attention line is between the inside address and the salutation. The salutation is then "Dear Sir:", "Dear Madam:", etc. The word "Attention:" is capitalized and followed by a colon and the addressee's job title, department, division, etc.
B. The enclosure is a reminder that the envelope contains more than the letter alone. The enclosure line is below the writer's typed name. The word "Enclosure:" is capitalized and followed by a colon and an identification of the enclosed material; such as, check, résumé, copy of receipt, etc.
C. The typist's initials, usually in lowercase and without periods, are used below the enclosure line when the person who typed the letter is not the person who wrote it.
D. The copy notation tells the addressee that a copy of the same letter was sent to another person. The notation is written "Copy:" capitalized and followed by a colon and an identification of the others to whom the letter was sent. It follows the typist's initials.

Note: If any of the last three special parts are unneeded, the below parts move up.

The following is an example of a full-block letter with all special parts:

Full-Block Letter with All Special Parts

7 Sampson Road
Lomita, California 90717
June 30, 2012

Laser Industries, Inc.
470 Madison Avenue
Auburn, California 95560

Attention: Personnel Manager

Dear Madam/Sir:

Since the addressee's name is unknown to the writer, the writer has used the "Attention:" line. The salutation, therefore, can't be the addressee's surname.

On the envelope, the mailing address and the attention line are used as they appear in the inside address above.

Sincerely yours,

Carla Springton

Carla Springton

Enclosures: application, transcripts, and résumé

dss

Copy: California Bureau of Employment Services

VIII. On the business letter envelope, use two parts: the return address and the outside address. The return address is in the upper left corner and has the writer's name (without a title of respect) on the first line, the writer's street address on the second line, and the writer's city, state, and zip code on the third line. The outside address is the same as the inside address in content and format. Fold the letter twice, resulting in three sections with a one-half inch shorter top division to aid in opening the letter.

The following is an example of an envelope with the addressee's name known to the writer:

```
James J. Johns                                              Postage
1339 Stafford Avenue                                        Stamp
Hickory, North Carolina 28602


                        Popular Science
                        Mr. Earl Williams, Electronics Editor
                        2 Park Avenue
                        New York, New York 10016
```

The following is an example envelope with the addressee's name unknown to the writer:

```
James J. Johns                                              Postage
1339 Stafford Avenue                                        Stamp
Hickory, North Carolina 28602


                        Popular Science
                        2 Park Avenue
                        New York, New York 10016

                        Attention: Electronics Editor
```

The following are examples of the modified-block style and the indented style of business letters.

Example of Modified-Block Style

> 7 Sampson Road
> Lomita, CA 90717
> June 30, 2012

Lazer Industries, Inc.
470 Madison Avenue
Auburn, California 95604

Attention: Personnel Manager

Dear Ms./Sir:

All content is the same in all styles of business letters.

This style makes two modifications to the full-block style. The heading's content ends at the right-hand margin, and the closing and signature are indented from the left margin the same distance as the heading.

All other content is arranged in the same format as the full-block style.

> Sincerely yours,
>
> *Carla Springton*
>
> Carla Springton

Enclosure: application form, transcript, and résumé.

dss

Copy: California Bureau of Employment Services

Example of Indented Style

>   7 Sampson Road
>   Lomita, CA 90717
>   June 30, 2012
>
> Lazer Industries, Inc.
> 470 Madison Avenue
> Auburn, CA 95604
>
> Attention: Personnel Manager
>
> Dear Ms./Sir:
>
>   The indented block style has only one change from the modified-block style. The first line of each paragraph is indented five spaces.
>
>   The content of all other parts is the same as the modified-block style.
>
>   Sincerely yours,
>
>   *Carla Springton*
>
>   Carla Springton
>
> Enclosure: application form, transcript, and résumé.
>
> dds
>
> Copy: California Bureau of Employment Services

The following illustrations show the folding and placement of the business letter into an envelope:

Step 1 : Fold the bottom 3 3/4" of the letter up.
Step 2 : Fold the top 3 1/2" of the letter down.
Step 3 : Place the folded letter into the envelope with the letter's free edge facing down.

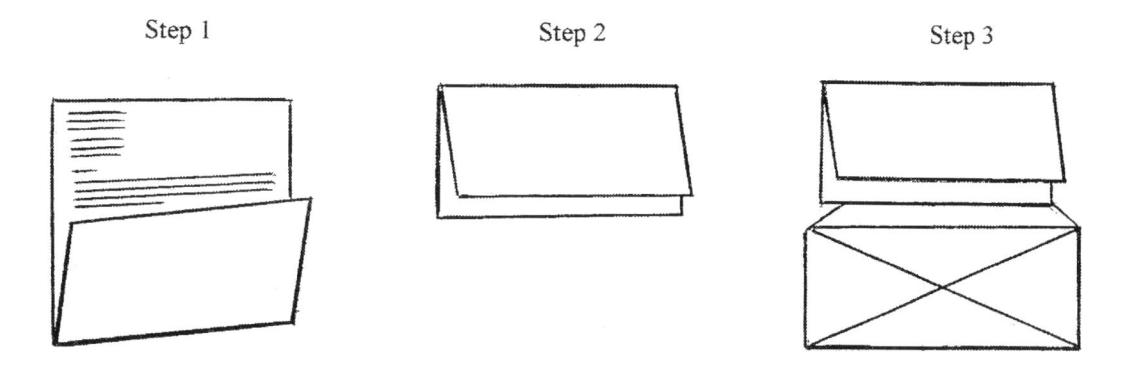

Bodies of E-Mails/Business Letters

The content depends upon the e-mail's or the letter's purpose, and each e-mail's or letter's purpose can be divided. Each division is usually a separate paragraph. The inquiry e-mail and all other e-mails would begin with the "Dear Sir / Ms.:"

The Letter of Inquiry (E-Mail)

The letter of inquiry asks for specific information; such as, a catalog, brochure, operating instructions, applications, etc. The body has three parts: specific, detailed identification of what is wanted, explanation of why the information is wanted, and a statement of appreciation for the addressee's help.

The following is an example of a sample inquiry e-mail:

Attention: Sales Department

Dear Sir:

Please e-mail me a copy of your firm's latest catalog and price list. I have just completed a three-level, 1500 square foot deck on my home, and I wish to furnish it with pressure-treated deck furniture from chairs to planters.
Thank you for answering my request.

Sincerely,

Thomas B. Wells

The Letter of Inquiry (Letter)

The following is an example of a full-block letter of inquiry:

1534 Summit Drive
Scranton, Pennsylvania 18505
March 2, 2012

Ronk's Pressure-Treated Deck Furniture, Co.
Route 30
Lancaster, Pennsylvania 17602

Attention: Sales Department

Dear Sir:

Please send me a copy of your firm's latest catalog and price list.

I have just completed a three-level, 1500 square foot deck on my home, and I wish to furnish it with pressure-treated deck furniture from chairs to planters.

Thank you for answering my request.

Sincerely,

Thomas B. Wells

Thomas B. Wells

The Order Letter (E-Mail)

Its body has four parts. The first paragraph is a specific identification of the ordering source (catalog or advertisement) and of the items wanted; including catalog number, manufacturer, material, size, shape, color, amount, unit price, etc. If ordering more than one type of item, make a chart to identify the items. The second part is an explanation of payment. The third part, if needed, is an explanation of desired shipment. The last paragraph is a one sentence statement of appreciation.

Cash should not be mailed. Credit cards, money orders, and personal checks are safer. Checks certified by a bank are helpful if the buyer wants to save the time needed for a personal check to "clear the bank." C.O.D. (Cash On Delivery) is sometimes an option. State sales taxes usually apply only to orders from within the same state.

Most businesses make it easy to order items online today. They provide order forms on their websites and allow for easy order of their products. In most cases it is as simple as filling in the form with the appropriate items and quantities. Normally, the predesigned order forms tabulate the cost of your order, add tax where necessary, and add shipping and handling charges automatically.

The following is an example of a sample order e-mail:

Catalog #	Description	Unit Price	Quantity	Total Price
FDC 100	Folding Deck Chairs	$69.95	4	$279.80
TDC 105	Two-Person Deck Chairs	$109.95	2	$219.90
RC 201	Rocking Chairs	$89.95	2	$179.90
PSS 301	Porch Swing and Supports	$189.95	1	$189.95
PTB 500	Picnic Table and Benches	$249.95	1	$249.95
HSP 600	Heart-Shaped Planters	$22.95	6	$137.70
FSP 601	Free-Standing Planters	$34.95	6	$209.70

Subtotal= $1,466.90
Tax 6% = $88.01
Shipping and Handling= $100.00
Grand Total= $1,654.91

Place credit card number here: _____

List expiration date of credit card: Month _____ Year _____

The following is an example of a full-block order letter:

1534 Summit Drive
Scranton, Pennsylvania 18505
April 4, 2012

Ronk's Pressure-Treated Deck Furniture, Co.
Route 30
Lancaster, Pennsylvania 17602

Attention: Mail-Order Department

Dear Sir:

Thank you for sending a copy of your 1996 catalog from which I want to order the following:

Catalog #	Description	Unit Price	Quantity	Total Price
FDC 100	Folding Deck Chairs	$69.95	4	$279.80
TDC 105	Two-Person Deck Chairs	$109.95	2	$219.90
RC 201	Rocking Chairs	$89.95	2	$179.90
PSS 301	Porch Swing and Supports	$189.95	1	$189.95
PTB 500	Picnic Table and Benches	$249.95	1	$249.95
HSP 600	Heart-Shaped Planters	$22.95	6	$137.70
FSP 601	Free-Standing Planters	$34.95	6	$209.70

Subtotal= $1,466.90
Tax 6% = $88.01
Shipping= $100.00
Grand Total= $1,654.91

Enclosed is a certified check for $1,654.91.

Please ship to the above address via freight.

Thank you for filling my order.

Sincerely yours,

Thomas B. Wells

Thomas B. Wells

Enclosure: certified check #67899 for $1,654.91

The Claim and Adjustment Letters (E-Mail)

The claim letter (from customer to company) tactfully requests the correction of a problem. Its body has four parts: an identification of the complete subject matter of original transaction, an explanation of the problem, a suggested correction, and a statement of appreciation.

The adjustment letter (from company to customer) explains the company's suggested correction of a problem. Its body has three parts: a summary of original transaction and problem, a correction for the problem, and a statement of appreciation.

The following is an example of a sample adjustment e-mail:

Attention: Mail-Order Department

Dear Sir:

Today I received the deck furnishings ordered from you on April 4, 2012, invoice #04671. The furniture is in perfect condition, but you included five HSP 600 Heart-Shaped Planters @ $22.95 each, one short of the six I ordered. You also included seven FSP 601 Free-Standing Planters @ $34.95, one more than ordered.

Since I can not substitute the extra planter for the omitted one, please inform me of how I can receive the additional Heart-Shaped Planter, return the Free-Standing Planter, and pay the additional charges.

Thank you for correcting my order.

Sincerely,
Thomas B. Wells

Enclosures: copy of original order letter of April 4, 2012
              copy of invoice #04671 for April 10, 2012

The following is an example of a full-block claim letter and an adjustment letter responding to the claim letter:

1534 Summit Drive
Scranton, Pennsylvania 18505
April 24, 2012

Ronk's Pressure-Treated Deck Furniture, Co.
Route 30
Lancaster, Pennsylvania 17602

Attention: Mail-Order Department

Dear Sir:

Today I received the deck furnishings ordered from you on April 4, 1996, invoice #04671. The furniture is in perfect condition, but you included five HSP 600 Heart-Shaped Planters @ $22.95 each, one short of the six I ordered. You also included seven FSP 601 Free-Standing Planters @ $34.95, one more than ordered.

Since I can not substitute the extra planter for the omitted one, please Inform me of how I can receive the additional Heart-Shaped Planter, return the Free-Standing Planter, and pay the additional charges.

Thank you for correcting my order.

Sincerely,

Thomas B. Wells

Thomas B. Wells

Enclosures: copy of original order letter of April 4, 2012
copy of invoice #04671 for April 10, 2012

Adjustment Letter Responding to a Claim Letter

Ronk's Pressure-Treated Deck Furniture, Co.
Route 30, Lancaster, Pennsylvania 17602
Telephone: (951) 633-0034

April 28, 2012

Mr. Thomas B. Wells
1534 Summit Drive
Scranton, Pennsylvania 18505

Dear Mr. Wells:

Thank you for calling our attention to our shipping error of April 10, 2012.

Tomorrow we will send your Heart-Shaped Planter HSP 600 at no additional charge. You may
also keep the extra Free-Standing Planter as a gesture of our apology for the delay.

We hope to serve you again.

Cordially yours,

James J. Ronk

James J. Ronk, Sales Manager

sm

The Sales Letter (E-Mail)

    The sales letter (from seller to prospective customer) is friendly, inviting, and informative. Its body includes four possible parts: an explanation of the product's or service's uniqueness (design, sizes, dependability, ease of operation, guarantee, appreciation, price, etc.), an explanation of the purchase procedure, possible offer of special inducements to buy now, and a statement of appreciation.

The following is an example of a full-block sales letter on letterhead stationery:

A & A Siding
"the best in the business"
42 Banister Avenue
Joplin, Missouri 64804
Phone : (417) 782-3900

June 1, 2012

Mr. and Mrs. David Sams
212 Oakmont Street
Cape Girardeau, Missouri 663701

Dear Mr. and Mrs. Sams:

A & A Siding has been providing courteous, prompt, inexpensive, and beautiful aluminum and vinyl siding installation for more than 20 years to more than 2000 satisfied customers.

Painting a home's exterior is expensive, hard, time consuming, and dangerous. Our clients do not worry about these—not ever again, guaranteed.

We deal in quantity, as well as quality, and have the best possible prices specially in the most popular designs, colors, and accessories.

If you are not interested in painting again, write or call us to make an appointment for a free estimate.

Very cordially yours,

John David Austin

John David Austin, President of A & A Siding

Mn

The Letter of Transmittal (E-Mail)

The letter of transmittal (also called the "cover letter" or "letter of application") introduces a job applicant and his or her résumé to a prospective employer. When a job resumé is submitted, it is the applicant's duty to know to include the letter of transmittal with the résumé, which is always an enclosure. This letter should always be an attachment to a brief e-mail. The e-mail simply notes the fact that the letter is attached.

This letter's body has five parts: an identification of the position sought and of how the applicant learned of it, a summary of the applicant's main qualifications, a referral to the enclosed résumé, a request for the next step in the employment process, and a statement of appreciation.

Following is the brief e-mail introduction to the attached résumé. I you know the person to whom you are writing use " Mr." or "Ms." followed by the person's last name.

Dear Ms. /Sir:

Please find the attached cover letter and résumé for your review. If you require additional
information, please contact me.

Thanks,

Todd Grossinger

The following is an example full-block letter of transmittal:

671 South Main Boulevard
Hohenwald, Tennessee 38462
May 12, 2012

M & Z Construction Company
Mr. Warren David, Personnel Manager
1430 Industrial Drive
Chattanooga, Tennessee 37422

Dear Mr. David:

Please consider me an applicant for the carpenter position advertised in the May 12, 2012, edition of the <u>Chattanooga Times</u> newspaper. I believe I can be an asset to your company. My education and experience will allow me to serve your company well.

On May 20th I will graduate from Johnson College in Scranton, Pennsylvania and receive an Associate in Applied Science degree. I am now seeking a position in my major, Carpentry and Cabinetmaking Technology.

Enclosed is my résumé for your review.

If you feel I qualified for the position, please contact me regarding the next step in your employment process.

Thank you for considering me for this position.

Respectfully yours,

Todd Grossinger

Todd Grossinger

Enclosure: résumé

Exercises for E-Mail/Business Letters

Exercise 1 : The Letter of Inquiry (E-Mail)

Directions : Rewrite the following letter of inquiry in full-block style. You will have to make changes in format, content, mechanics, and syntax. address an envelope example for the letter. or Change this letter to an e-mail.

Michael M. Montana
514 Glenside Ave., E. Hardford, Connecticut 06108
July 25 2112

Department M/O
Hydem Jeep®Accessories Co
Box 407
Kensington, Kansas, 66951
Telephone: (913) 476-2241

Dear sir,

    I'm writting this letter to you concerning having a catalog for free that I saw advertised in a magazine a while ago. I bought a newJeep®. I want more parts for it. I'm real enthused about getting hard top to use in place of the summer soft top. During the months of Oct. through Apr.

Send the catalog to the home address of Michael M. Montana 514 Glensid Ave., E. Hardford Connecticut, 06108.

                                                  Thank you
                                                  Michael M. Montana

Exercise 2 : The Letter of Inquiry (E-Mail)

Directions : Choose a topic and write a full-block letter of inquiry or an e-mail. Make up addressee's name or attention line and address if necessary.

1. Ask a major car manufacturer how many models of a car exist.
2. Assuming that you own a retail store, request information from a manufacturer whose product line you want to carry.
3. Ask a famous person (athlete, politician, entertainer, etc.) if he or she would autograph an item received from you.

Exercise 3 : The Order Letter

Directions : Choose a topic and write a full-block order letter purchasing at least three different items. Make up addressee's name or attention line, address, and products' complete descriptions. Include shipping charges and your state's sales tax.

1. Order home windows and/or doors.
2. Order bathroom furniture and/or fixtures.
3. Order tools used in your major.
4. Order auto, truck, motorcycle, etc., accessories.
5. Order kitchen appliances.

Exercise 5 : The Claim and Adjustment Letters (Email)

Directions : Assume that one item of your previous order letter was received in damaged or incomplete condition. Write a full-block style letter or e-mail to inform the seller of the problem.

Directions : Assume that you are the seller of the damaged or incomplete item noted in the claim letter. Write a full-block style letter or e-mail to inform the buyer of the adjustment.

Exercise 6 : The Sales Letter

Directions : Assume that you own a company that sells, services, or repairs. Write a full-block style sales letter or an e-mail to a possible customer. Make up the customer's name and address, as well as your company's product or service.

Exercise 7 : The Sales Letter

Directions : Suppose that you own a business that sells through its catalog. Compose an order form to accompany the catalog. Include your company's identification, the order's identification, the description of purchase chart (including catalog #'s, brand, quantity, size, price, etc.), shipment, payment, etc.

Exercise 8 : The Letter of Transmittal

Directions : Assume that you will graduate from your college shortly. Write a full-block style letter of transmittal answering the following ad from today's edition of your local newspaper.

**Now Hiring Tradesmen and Technicians
Carpenters, Electricians, Auto Technicians, Draftspersons, Precision Machinists, Radiologic Technicians, Welders, Veterinary Technicians,
Biomedical Technicians, HVAC Technicians, or ? (your major)**

Requirements: Experience and/or education

Send résumé to American Technical Company, Inc.
720 E. Chestnut Street
Chicago, Illinois 60611
Attn: Mr. Jay Walters, Employment Manager

Chapter XII : The Personal Résumé

The personal résumé is a form of a job application created by the applicant rather than by the employer. The résumé is slanted to show the applicant's best employment qualities. The word "personal" in reference to "résumé" means not only that the résumé is about the applicant but also that it is created by the applicant. A résumé writing service can be compared to having an impostor play the part of the applicant at an interview. The résumé should be completed early in one's job search, updated periodically, and sent immediately upon learning of a job possibility. A personal résumé is a constant work in progress. The format and contents of résumés change constantly. It is important to keep up-to-date with trends in formatting and contents if the job seeker is to be successful in his/her job search.

The first step in preparing a résumé is to complete a personal inventory as listed below. This inventory is also helpful in completing an employment application and answering questions at a job interview.

The Personal Inventory

Directions : Please complete this inventory and update it as needed. When appropriate, explain your answers. Add paper if necessary.

Identification of Applicant

Full Name :_____

Street Address : _____

City, State, and Zip Code : _____

Phone Number and Area Code :   Home ( ) _____
                                Work ( ) _____

E-mail Address : _____

Career Objectives

Immediate : _____

Long Range (after 5, 10 , etc. years) : _____

_____

Personal Information

Age (and date of birth) : _____

Appearance (height and weight) : _____

Marital Status (and dependents) : _____

Health (work limitations and last physical) : _____

Citizenship (and Social Security number) : _____

Hobbies: _____

Community Activities : _____

Licenses or Certifications : _____

Special Skills : _____

Education
(post secondary, college, trade school, etc.)

Name of College, Trade School, Etc. : _____

Street Address : _____

City, State, Zip Code : _____

Dates of Attendance : _____

Major Study and Its Description : _____

_____

_____

Additional Studies : _____

_____

Favorite Course : _____

Favorite Projects : _____

Toured : _____

Seminars Attended : _____

Most Challenging Course : _____

Most Improved Course : _____

Attendance : _____

Extracurricular Activities : _____

Awards and Honors : _____

Degree, Diploma, Etc. : _____

(High School)

Name of High School : _____

Street Address : _____

City, State, and Zip Code : _____

E-mail Address : _____

Dates of Attendance : _____

Major Study : _____

Additional Studies : _____

Favorite Course : _____

Most Challenging Course : _____

Most Improved Course : _____

Attendance : _____

Extracurricular Activities : _____

Awards and Honors : _____

Type of Diploma : _____

Work Experience (List most to least recent)
Most Recent Employment

Company Name : _____

Street Address : _____

City, State, and Zip Code : _____

Telephone Number : _____

E-mail Address : _____

Dates of Employment : _____

Weekly Hours : _____

Duties : _____

Wages: _____

Supervisor(s) : _____

Reason for Leaving : _____

Second Most Recent Employment

Company Name : _____

Street Address : _____

City, State, Zip Code : _____

Telephone Number : _____

E-mail Address : _____

Dates of Employment : _____

Weekly Hours : _____

Duties : _____

Wages : _____

Supervisor(s) : _____

Reason for Leaving : _____

Third Most Recent Employment

Company Name : _____

Street Address : _____

City, State, Zip Code : _____

Telephone Number : _____

E-mail Address : _____

Dates of Employment : _____

Weekly Hours : _____

Duties : _____

Wages : _____

Supervisor(s) : _____

Reason for Leaving : _____

Fourth Most Recent Employment

Company Name : _____

Street Address : _____

City, State, Zip Code : _____

Telephone Number : _____

E-mail Address : _____

Dates of Employment : _____

Weekly Hours : _____

Duties : _____

Wages : _____

Supervisor(s) : _____

Reason for Leaving : _____

Fifth Most Recent Employment

Company Name : _____

Street Address : _____

City, State, Zip Code : _____

Telephone Number : _____

E-mail Address : _____

Dates of Employment : _____

Weekly Hours : _____

Supervisor(s) : _____

Reason for Leaving : _____

Military Experience

Branch : _____

Dates : _____

Locations and Duties : _____

_____

Schooling : _____

_____

Awards and Decorations : _____

_____

Rank at Discharge : _____

Type of Discharge : _____

References (See note below.)

Name and Position : _____

Street Address : _____

City, State, and Zip Code : _____

Telephone Number : _____

E-mail Address : _____

Name and Position : _____

Street Address : _____

City, State, and Zip Code : _____

Telephone Number : _____

E-mail Address : _____

Name and Position : _____

Street Address : _____

City, State, and Zip Code : _____

Telephone Number : _____

E-mail Address : _____

Name and Position : _____

Street Address : _____

City, State, and Zip Code : _____

Telephone Number : _____

E-mail Address : _____

Name and Position: _____

Street Address : _____

City, State, and Zip Code : _____

Telephone Number : _____

E-mail Address : _____

Note: Five references are sometimes required. They should not be relatives. They should be people who know you well. They are sometimes classified as "personal references," your friends, and "professional references," those whom you have worked for or studied under. All references should be contacted prior to using their names on a résumé. All addresses and contact information should be verified by the references.

Formats for Résumés

After completing the personal inventory, turn it into a résumé by listing its information that describes you best. The format is different for everyone, but the similar contents are listed in chronological order from personal identification to references. The categories identifying the content of the personal inventory are not on the résumé; they are understood. The category of "Personal Information" should not be on the résumé. The category of "Military Experience" can be omitted but should be included when applicable for the position sought..

Résumé Hints

1. Use quality, unlined, white, 8 1/2" X 11" business-letter stationery. Quality cotton bond paper and matching envelopes are available for job seeking. This paper and envelopes makes a better impression when job seeking.
2. Allow at least a 1" margin around the page.
3. Limit your résumé to one typed page.
4. References should be listed on a second page using the same heading as found on the first page of the résumé. Do not forward the reference page to an employer until asked to do so.
5. List education and work experience from most to least recent.
6. Be consistent in the spacing, listing, and content of all categories.
7. Check for correct spelling.
8. Be complete in zip codes, street addresses, telephone numbers, etc.
9. Remember that the résumé is always an enclosure with the letter of transmittal.
10. Avoid too much white space on the résumé. Try to balance the content on the page as a picture in a frame. Use both the left and right margins as the border for the résumé.
11. Keep current with résumé trends through reading up-to-date websites and articles. The résumé format must be current to grab the attention of the personnel department.
12. Avoid using predesigned templates. If everyone used these, the résumés would look like an application form. Keep the résumé personal by creating your own distinctive format.
13. Always keep the résumé professional.
14. Highlight the skills and education that fit the advertised job description.

On the following pages are sample résumés showing a range of format styles.

Jonathan A. Brown

520 Court Street, Apartment 12
West Allis, Wisconsin 53227
Telephone : 414.327.7440
E-mail : jbrown@hotmail.com

Objective

To secure an entry-level position as a carpenter (willing to relocate)

Education

8/ 2010 to 5/ 2012	Allis Trade School 1414 South Seventh Street Allis, Wisconsin 53202	Major : Carpentry Diploma : Certificate of Completion (B+ average)
9/ 2003 to 6/ 2007	West Allis High School/ Allis Vo-Tech 400 Main Avenue West Allis, Wisconsin 53227	Vo-Tech Major : Masonry Diploma : Vocational

Work History

7/ 2010 To Present	Masonry Effects 78 West Irving Street Milwaukee, Wisconsin 53227	Title: Mason's helper Duties: Set brick, concrete blocks, cinder blocks; installed stucco; finished concrete
Summer of 2003 & full-time 2004-2005	D. J. Constructions in Stone 1612 Ridge Street Milwaukee, Wisconsin 53207	Title: Laborer Duties : Mixed mortar, set up scaffolds on construction sites
7/ 2003 To 9/ 2005	McDonalds Restaurant 17 State Street Milwaukee, Wisconsin 53207	Title: Short-Order Cook Duties: Operated fryer, grill

References

Furnished upon request

Jonathan A. Brown

520 Court Street, Apartment 12
West Allis, Wisconsin 53227
Telephone : 414.327.7440
E-mail : jbrown@hotmail.com

References

Mr. Daniel Jones, Owner of D. J. Construction
1612 Ridge Street
Milwaukee, Wisconsin 53207 Telephone : 414.769.7071
             E-Mail : DJConstruction@aol.com

Mr. Gregory Samsell, Foreman at D.J. Construction
1612 Bartley Avenue
Milwaukee, Wisconsin 53207 Telephone : 491.769.4433
             E-Mail : gsamsell@yahoo.com

Mr. Joseph J. Grant Jr., Carpentry Instructor
Allis Trade School
1414 South Seventh Street   Telephone : 414.327.6505
Allis, Wisconsin 53202   E-mail: jgrant@allis.edu

Mr. James Smather, Owner of Masonry Effects
2456 West Lemon Avenue
Milwaukee, Wisconsin 53201 Telephone: 414.735.3334
             E-Mail: jsmathermasonry@aol.com

Jennifer L. Lennox

320 West Madison Avenue, Irwindale, California 91702
Telephone : 818.696.0634   E-Mail : jlennox@aol.com

Employment Objective

Architectural draftsperson with the chance for advancement in your progressive company

Educational Background

Napa Junior College (9/2010-5/2012)
100 College Road, Nappa, California 94559
Major : Architectural Drafting and Design
Internship : Grant's Design and Landscaping (3 months)
Activities : Drafting Club, Volleyball, and Softball
Honors : Honor Roll (3 semesters, 3.22 GPA), captain of softball team 2010, vice-president of Drafting Club 2010
Certification : California Architectural Drafting Certificate
Degree : Associate of Arts

Irwindale High School (9/2006-6/2010)
1670 Boulevard Avenue, Irwindale, California 91702
Activities : Volleyball (1-4), Softball (1-4), Student Government (3-4), Yearbook (4)
Honors : Honor Roll (4 yrs.), captained the volleyball team and named to league all-star volleyball team 2009
Diploma : Academic

Employment Background

McDonald's Restaurant (10/2008-Present)
1600 O'Neill Highway, Irwindale, California 91702
Responsibilities : Cook and wait on counter and drive-through customers for three years, supervised the 4-11 shift for the past seven years.

Irwindale Post (7/2006-5/2008)
150 Downtown Square, Irwindale, California 91702
Responsibilities : paper carrier for 40 to 70 customers

Jennifer L. Lennox

320 West Madison Avenue, Irwindale, California 91702

Telephone : 818.696.0634   E-Mail : jlennox@aol.com

References

Miss Paula Wallenta, Retired Registered Nurse   Telephone : 714.661.1871
332 West Mason Lane, Laguna Niguel, California 92697

Mr. Robert Reinheimer, Owner of Mc Donald's Franchise
1600 O'Neill Highway, Irwindale, California 91702   Telephone : 818.969.7270

Mr. Christopher Gardier, Manager of Mc Donald's
816 Quincy Avenue, Irwindale, California 91702   Telephone : 818.969.7061

Patrick Thomas Reed

611 Webster Avenue, Midlothian, Texas 76065
Telephone : 214.723.6071

Career Objective

To secure a position in your biomedical department and continue my education to advance in my profession

Education

Midlothian Trade and Technical Institute (2010 to 2012)
Center Street Campus, Midlothian, Texas 76065
Major: Biomedical Technology
Course Description : Seventy percent of curriculum was theory and practice in safety, testing, installing, servicing, monitoring, and calibrating hospital equipment. Thirty percent was academic studies of English, math, blueprint reading and drafting, and business.
Honors : Dean's List (4 semesters, 3.82 GPA)
Degree : Associate in Applied Technology

Military

US Army (2002 to 2008)
Locations : Fort Dix, New Jersey; Fort Brag, North Carolina; Hamburg, Germany
Schools : Electronic Surveillance, Military Communications, and Radar
Duties : Mobil Radar Specialist, responsible for the set up, operation, service, and disassembly of transportable radar installations
Rank and Discharge : Sergeant, honorable, current member of Radar Detachment of the 107th Infantry Army Reserves

Work

Boggs Machine Shop, Inc. (2010 to present)
21 Badger Road, Midlothian, Texas 76065
Duties: Part-time (20 hrs. per week) electronic maintenance man

C & W Electronic Controls (2008 to 2010)
4228 Commercial Drive, Irving, Texas 75061
Duties : Installed, serviced, and repaired electronic machine controls for five years Worked as troubleshooter, estimator, and salesman of electric controls for three years. Served as crew chief for calibration of large, assembly-line electric controls, motors, and computers for last three years with company.
Reason for Leaving: Bankruptcy of Company

Exercises for the Résumé

Exercise 1 : Personal Résumé

Directions : Compose a one-page, up-to-date personal résumé. Compose a second page containing five references. Use an example résumé format or create your own.